A. Schneeweiss M. Weiss

Advances in Nitrate Therapy

Second, Revised and Enlarged Edition

Springer-Verlag
Berlin Heidelberg New York London
Paris Tokyo Hong Kong Barcelona

Adam Schneeweiss, MD
9, Kehilat Sofia Street
Tel Aviv 69018, Israel

Marija Weiss, DSc
Arenzhofstrasse 28
5000 Köln 71, FRG

ISBN-13: 978-3-642-75836-2 e-ISBN-13: 978-3-642-75834-8
DOI: 10.1007/978-3-642-75834-8

Library of Congress Cataloging-in-Publication Data
Schneeweiss, Adam. Advances in nitrate therapy / A. Schneeweiss, M. Weiss. –
2nd rev. and enl. ed. Includes bibliographical references.
ISBN 0-387-52798-2 1. Nitrates – Therapeutic use. 2. Angina pectoris –
Chemotherapy. I. Weiss, M. (Marija) II. Title. RC685.A6S36 1990
616.1'22061 – dc20

2127/3140 – 543210 – Printed on acid-free paper

Contents

A. Introduction and Mechanism

1. Introduction

Nitrates have been for many years the cornerstone of cardiovascular therapy for various indications. Not only have nitrates stood the test of time for treatment of chronic stable angina pectoris, but the indications for them have markedly expanded. They now include all forms of angina pectoris and myocardial ischemia, congestive heart failure and hypertensive emergencies. The beneficial effects of nitrates in all these conditions result from their vasodilatory properties, but it is still unclear whether the central or peripheral effects predominate in the therapeutic mechanism.

Recently nitrates have been shown to fulfill the most important requirement for each and every drug - to reduce mortality. A large scale study revealed that isosorbide dinitrate, combined with hydralazine, reduced long-term mortality by 28% in patients with congestive heart failure. This finding will certainly stimulate research on nitrates, and it might be expected that interest in these drugs will markedly increase.

The primarily used nitrates in clinical practice are nitroglycerin and the complex organic nitrates, mainly isosorbide dinitrate. The organic mononitrates are now under clinical investigation, and isosorbide-5-mononitrate (IS-5-MN) appears to be especially promising.

All investigators agree upon the short-term efficacy of nitrates in relief of angina pectoris and hemodynamic disturbances. Some investigators, however, claim that long-term administration of nitrates is associated with the development of tolerance. This claim is not only controversial scientifically but has not stood the test of clinical practice. Every clinician can tell from his own experience that nitrates are effective over many years of treatment. Certain techniques may, however, improve the long-term efficacy of nitrates. These include the use of low doses, eccentric dosing and nitrate-free intervals.

The expanding indications for nitrates and the increasing number of those receiving them have stimulated efforts to improve and simplify delivery of these agents. Among the most promising new delivery systems are the transdermal devices which supply a constant plasma level of nitrates throughout 24 h. It has been claimed that this constant level may be associated with early development of tolerance. One of the solutions suggested is treatment for about 16 h during daytime, with removal of the transdermal devices during the night. However, the possibility of tolerance developing to transdermally administered nitrates is still controversial, as it has been reported for only some of the available formulations.

Another new and important delivery system is the oral spray of isosorbide dinitrate (ISDN) or nitroglycerin. The most rapid conventional route of administration of nitrates to ambulatory patients is sublingual administration of tablets. Absorption of nitrates starts after dissolution of the tablet – a process which takes time and limits the rapidity of effect. The oral spray avoids this delay and also allows more rapid and convenient administration by medical personnel. Other exciting fields of investigation are the mechanism of action at the molecular level and new indications such as secondary prevention after acute myocardial infarction.

An important new line of research is differential selectivity of the various nitrates. Generalized vasodilation is not always optimal, and in certain important indications selective vasodilation is preferred. For example, in patients with ischemic heart disease and angina pectoris or acute myocardial infarction venodilation and coronary dilation are definitely beneficial. However, arteriolar dilation might be deleterious if it results in an excessive reduction of coronary perfusion pressure. Hence those nitrates which preferentially dilate the venous more than the arterial bed might be superior to other nitrates in these conditions. Preliminary data suggest that IS-5-MN might have such a therapeutic profile. The same rationale also applies to congestive heart failure, where venodilation is now considered to be the primary therapeutic effect.

Thus about 100 years after the use of nitrates began, they are not only among the most important cardioactive agents, but also a group of agents undergoing dynamic and continuous investigation.

It is the purpose of this book to describe the state of the art in the field, to report current fields of progress and to indicate some points which will require research in the future. Nitroglycerin, isosorbide dinitrate and isosorbide mononitrate are discussed in detail. Some new and investigational nitrates are briefly reviewed.

1.1. History

Nitroglycerin was the first nitrate to be used in therapy. It was synthesized by Sobreto in 1846. Sublingual nitroglycerin has been available for about 140 years. Even in its early days it was noted that sublingual nitroglycerin can cause severe headache. Soon after its synthesis, attempts were made to use nitroglycerin for angina pectoris, with little success. Amyl nitrite was then considered to be more effective than nitroglycerin for this indication. However, the problems inherent in the use of amyl nitrite, including the unpredictable dose-response relationship and the very short duration of action, resulted in a search for other agents and increased interest in nitroglycerin. About 100 years ago sublingual nitroglycerin was established as a standard agent for relief of pain in patients with angina pectoris. In those early days it was also considered effective for angina prophylaxis.

The other organic nitrates such as ISDN, erythrityl tetranitrate, pentaerythritol tetranitrate and manitol hexanitrate were developed and introduced into clinical practice when it was realized that the effect of sublingual nitroglycerin is short-lasting and that the effect of oral nitroglycerin is limited and unpredictable. Later, when the pharmacokinetic profile of the various nitrates was better understood and the therapeutic requirements in different cardiovascular diseases were better defined, new agents, such as the mononitrates, were developed and studied.

All these agents are nitrate esters or nitrite esters, characterized by a sequence of carbon-oxygen-nitrogen, which is essential for vasodilator properties. Therefore nitroglycerin is actually not a nitro compound and the name glyceryl trinitrate may be preferred. The organic nitrites, such as amyl nitrite, are esters of nitrous acid (NO), and the organic nitrates are esters of nitric acid (NO_2).

The use of organic nitrates in prophylactic clinical practice has largely been delayed by a conclusion made by a group of investigators and reported first in 1972 and repeated thereafter. Needleman and coworkers [1] reported that orally administered organic nitrates were ineffective because of complete first-pass hepatic metabolism. Based on this thesis the use of long-acting organic nitrates declined markedly. It took almost a decade until the pharmacokinetic and pharmacodynamic profiles of ISDN and its metabolites were fully understood. Only then was it possible for Needleman's thesis to be rejected on a strong scientific basis.

2. The Mechanism of Action of Nitrates at the Cellular Level

Nitrates are considered "direct-acting" vasodilators. Their effect is not mediated by a neurohumoral mechanism involving the adrenergic system, the renin-angiotensin system or other vasodilatory mechanisms. Although it has been suggested that some of the dilatory effect of nitrates is exerted by the release of vasodilating prostaglandins, most investigators believe that this is not the main mechanism of action. Thus it is generally accepted that nitrates act directly, at the cellular level, in vascular smooth muscle.

The fundamental biochemical mechanisms of action of nitrates have been extensively studied in the past two decades. Almost all studies have been performed with nitroglycerin. One of the most important achievements in the field was the study by Needleman et al [2], who observed that incubation of aortic strips with ethacrynic acid, which is a sulfhydryl alkylating agent, resulted in a decrease in sensitivity to nitroglycerin. These investigators also suggested that tolerance to nitroglycerin could be induced by oxidation of sulfhydryl groups with the formation of bisulfide [3]. These findings support the view that the direct vasodilator action of nitroglycerin is mediated by sulfhydryl groups on the membrane of vascular smooth muscle.

According to this theory the sulfhydryl group is the essential component of receptors to nitrates in cell membrane. Interaction between nitrates and sulfhydryl groups results in reversible oxidation of these groups, with formation of a disulfide group. This group has a low affinity for nitrates. Interaction of nitroglycerin with the sulfhydryl groups forms 5-nitrosothiol groups, which activate the enzyme guanylate cyclase [4]. Activation of this enzyme results in an increase in intracellular concentrations of guanosine 3-monophosphate (cyclic GMP). This has recently been shown in aortic preparations, in vivo and in vitro with nitroglycerin, ISDN and even IS-5-MN.

Recently Horowitz et al [5] reported a study evaluating whether the availability of sulfhydryl groups modulates responsiveness to nitroglycerin. As S-nitrosocysteine is the S-nitrosothiol most effective in stimulating guanylate cyclase activation, the investigators evaluated whether the availability of this compound affects the hemodynamic response to nitroglycerin in patients undergoing cardiac catheterization. The hemodynamic response was measured before and after infusion of N-acetylcysteine, which is a source of cysteine. The N-acetylcysteine potentiated the vasodilator effect of nitroglycerin. This finding supports the theory that the availability of sulfhydryl groups determines the responsiveness to nitroglycerin.

Ignarro and Gruetter [4] have shown that inhibition of guanylate cyclase by methylene blue attenuates the vasodilator effect of nitrates.

It was suggested that individual differences in hemodynamic response to nitrates result from variations in availability or oxidative state of sulfhydryl groups in membranes of vascular smooth muscle.

The rise in cytosolic cGMP subsequently results in phosphorylation of target proteins by cGMP-dependent protein kinase. cGMP-dependent protein kinase can relax vascular smooth muscle cells which have been chemically treated to damage their cell membrane and, therefore, have no functioning cell membrane. This finding indicates that the contractile proteins themselves might be one of the targets of the cGMP-dependent protein kinase activated by nitrates.

The increase in cytosolic cGMP results in lowering of cytosolic calcium level, leading to smooth muscle relaxation. This lowering of cytosolic calcium may occur via cGMP-dependent activation of the sarcolemmic calcium extrusion ATPase.

Another theory explains the mechanism of vasodilation by the release of prostaglandins induced by the interaction of nitrates and receptors. Nitrates were shown to increase production of prostacyclin, the most potent vasodilator prostaglandin, in endothelial cells from umbilical veins [6]. Nitroglycerin was shown to increase the production of prostaglandins by the cardiac vasculature, as evidenced by an increase in prostaglandin E levels in the coronary sinus after infusion of nitroglycerin [7]. These effects may be mediated by stimulation of cyclooxygenase or by inhibition of thromboxane synthetase.

However, at least one study revealed that pretreatment with indomethacin, an inhibitor of prostacyclin synthesis, did not attenuate the antianginal effect of nitroglycerin [8]. Inhibition of cyclooxygenase also failed to attenuate the increase in renal blood flow produced by nitroglycerin [9]. In several isolated arterial preparations nitroglycerin did not increase the levels of prostaglandins [10]. All these findings indicate that the vasodilator effect of nitrates is not mediated by prostaglandins.

Another important argument against a potential role of prostaglandins in the mechanism of action of nitrates is the finding that no increase in urinary excretion of metabolites of prostaglandins has been observed in human subjects after treatment with nitrates. Although this does not preclude a local effect on prostaglandins in vascular walls, most investigators now believe that the effect of nitrates is not mediated by prostaglandins.

2.1. Metabolism of Nitrates in Vascular Walls

Only recently has it been realized that nitrates are also metabolized (denitrated) in the vascular wall and that this process may be, quantitatively, more important in the elimination of nitrates than hepatic metabolism. Moreover, vascular wall metabolism is most probably directly linked to the mechanism of action of nitrates. This is discussed further in the chapter on the pharmacokinetics of nitrates.

2.2. *Endothelium-derived Relaxing Factor (EDRF): The Endogenous Nitrate*

The mechanism of action of nitrates and the character and activity of the endothelium-derived relaxing factor (EDRF) have been investigated independently until it has been recently realized that they are closely related, functionally and perhaps also chemically.

Nitrates act through stimulation of soluble guanylate cyclase, resulting in elevation of cyclic GMP levels. Nitrates can release nitric oxide (NO) and, as NO can by itself activate soluble guanylate cyclase [10a] it is likely that nitrates act through release of NO. The chain of reactions involves an interaction with thio-amino acids to form unstable nitrosothiol compounds which can release NO groups. NO can directly relax vascular smooth muscle.

In parallel with these findings it has been shown that certain vasodilating stimulations act through or are enhanced by the intact vascular endothelium cells. Ten years ago a report on "the obligatory role of endothelial cells in the relaxation of smooth muscle by acetylcholine" has been published [10b]. Studies by Ignarro et al [10c, 10d] and by others demonstrated the relationship between the EDRF and accumulation of cyclic GMP. The EDRF was shown to be different from the natural vasodilator prostacyclin. The accumulating data indicating not only a functional but also a chemical resemblance between EDRF and NO resulted in the suggestion that they are the same or closely related [10e, 10f, 10g]. Description of the detailed chemical argumentation supporting this hypothesis is beyond the scope of this book but there is direct evidence that both arterial and venous EDRF is related to NO.

It seems, therefore, that the active group of organic nitrates is similar to one of the most important natural vasodilator, EDRF. Nitrates may serve as a precursor supplying NO when required. This interaction with natural vasodilation is an important physiologic mechanism of nitrate activity. EDRF may now be termed "endogenous nitrate".

B. Indications

3. Stable Angina Pectoris

Nitrates are the cornerstone in treatment of anginal episodes of all types and, in most centers, also in prevention of such episodes. The sublingual or intravenous administration of nitrates for relief of pain in patients with an acute anginal episode is universal. The use of long-acting oral nitrates, in high doses around the clock, is recommended by most, but not all, cardiologists for prevention of anginal episodes. Those concerned with tolerance use high doses at long time intervals. Others use very low doses. In a few centers nitrates are not used at all for prophylaxis but only for treatment of anginal episodes. Several studies have now been performed to resolve these questions and to standardize treatment.

Nitrates have been found effective in all forms of angina. Their anti-ischemic effect has been found beneficial in chronic stable angina pectoris, unstable angina, vasospastic angina, postinfarction angina, angina associated with hypertension or congestive heart failure and even in some of the patients with angina and normal coronary arteries. They are effective in all age groups. All nitrates exert an antianginal and anti-ischemic effect. Pharmacokinetic differences and, recently, also some differences in hemodynamic profile indicate the selection of specific nitrates for various diseases and patients.

Nitroglycerin, the first drug used in treating angina pectoris, is still used today for this indication. It is usually the first line of treatment. It rapidly relieves pain in most patients with classical symptoms of chronic stable angina pectoris and in many patients with intermediate coronary syndromes. In addition, it decreases the electrocardiographic changes of ischemia (ST-segment depression or elevation) and increases exercise capacity and duration.

Orally administered nitroglycerin is poorly effective because of the extensive first-pass hepatic metabolism (see section on clinical pharmacology). The intravenous route is reserved for hospitalized patients with severe refractory angina. In the majority of ambulatory (and hospitalized) patients with angina pectoris sublingual administration of nitroglycerin is the most practical and rapid route of administration. Recently it has been shown, however, that ISDN oral spray acts more rapidly than sublingual tablets. This topic is further discussed in the chapter on nitrate oral spray. These routes avoid the first-pass hepatic metabolism.

Sublingual nitroglycerin tablets are effective within 1–5 min and the effect is maintained for 15–30 min. These parameters determine the clinical profile of sublingual nitroglycerin use for relief of acute pain and for prevention of pain before a short effort, which is known to induce chest pain.The rapid elimination of nitroglycerin makes it ineffective in preventing pain over long periods. On the other hand, it does not achieve a constantly high plasma level and, therefore, does not induce tolerance. The only practical methods of using nitroglycerin to prevent pain over long periods are transdermal or buccal administration. These methods are discussed in a separate chapter.

Other organic nitrates are also given sublingually to relieve pain. ISDN is most widely used for this indication. It is effective within 1–5 min from sublingual administration, and its effect is sustained for longer periods than that of nitroglycerin. Therefore it may be used also for prevention of pain before a prolonged effort (30–90 min) which is known to cause chest pain. ISDN oral spray acts more rapidly than sublingual tablets of ISDN and is used for the same indications.

Similarly to nitroglycerin, ISDN does not only relieve pain, but also diminishes electrocardiographic evidence of ischemia and increases exercise tolerance in patients with angina pectoris. ISDN also undergoes extensive first-pass hepatic metabolism, but unlike nitroglycerin considerable amounts of the drug reach the systemic circulation. Therefore it may be given orally. By this route of administration ISDN exerts its antianginal effect within 10–30 min. Because ISDN has a longer duration of action than nitroglycerin, it is used to prevent pain over 24 h. However, the regimens required to achieve this goal are considerably different in various centers. This is mainly due to the controversy about the duration of action of oral ISDN and about possible attenuation of the effect and/or shortening of the duration of action with chronic treatment. Therefore the dosage intervals of orally administered ISDN range from 2–12 h. The interval depends, of course, also on the severity of the disease. Most patients receive oral ISDN three to four times daily. The doses used for improvement in exercise tolerance may be higher than those required for relief of pain.

The antianginal effect of nitrates is associated with a decrease in the rate-pressure product at rest as a result of a reduction in systolic arterial pressure. Heart rate is not altered or is slightly increased. Conflicting data on the effect of ISDN on the rate-pressure product at the onset of exercise-induced angina have been reported. Some investigators have reported no change in the rate-pressure product at the onset of angina, whereas others have found an increase in the product [11, 12]. Others reported an increase during acute therapy and no change during sustained therapy [13]. Left ventricular regional function has improved during nitrate therapy in patients with coronary artery disease and angina pectoris [14].

Thus, the various formulations of nitroglycerin and other nitrates cover the whole spectrum of therapeutic requirements in angina pectoris. Their therapeutic capabilities have been improved by development of new delivery systems, new formulations (such as mononitrates, discussed in a separate chapter) and new dosage regimens.

Intravenously administered ISDN was found to be very effective in patients with severe angina pectoris refractory to other drugs as well as to nitrates administered by other routes of administration. In extreme cases, especially those associated with coronary spasm, intracoronary infusion of nitrates may be the only effective therapy.

Recently IS-5-MN, the long-acting active metabolite of ISDN, has been found effective in long-term prevention of anginal episodes. It produces the whole spectrum of anti-ischemic effects of nitrates and might have the advantage of exerting this effect without marked reduction in systemic vascular resistance and coronary perfusion pressure.

4. The Mechanism of Action of Nitrates in Angina Pectoris

Although nitrates have been used for 100 years in angina pectoris, the *accurate* mechanism of their antianginal effect is not clear. This is not because the mechanism of their effect is unknown. On the contrary, several mechanisms, each of which contributes to the effect, have been described. However, the relative importance of these mechanisms is not known, and it is therefore difficult to specify the predominant mechanism.

The suggested mechanisms involve decrease of myocardial oxygen demand (by preload and afterload reduction), improvement in distribution of coronary flow (by reduction of left ventricular filling pressure and opening of coronary collaterals), and increased coronary supply (by coronary dilation and relief of spasm).

4.1. Coronary Dilation

It is reasonable to assume that coronary dilation is the mechanism of antianginal effect of nitrates, as myocardial ischemia results from obstruction of coronary arteries and nitrates are potent dilators of coronary arteries.

The situation is, however, highly complex. Dilatory effects of nitrates have been demonstrated in various segments of the coronary circulation: epicardial (diseased and normal) coronary arteries, coronary resistance arterioles, coronary collaterals and coronary veins. Not all of these effects are always beneficial. Moreover, the relative extent and time course of the vasodilatory effect in different segments of the circulation are important in determining the therapeutic effect. These parameters were shown to vary in different coronary regions.

If we add to this multifactorial situation the interaction between coronary and systemic mechanisms and the variability of mechanisms responsible for myocardial ischemia in the different conditions associated with it, the complexity of the mechanism of action of nitrates may be appreciated. This is one of the reasons why after 100 years of use in angina pectoris their mechanism of action is still controversial.

It was shown that nitroglycerin, infused directly into the coronary arteries at doses too low to exert a systemic effect, dilates the coronary arteries. The drug has a preferential effect on the large coronary arteries. For example, Feldman et al [15] reported that several doses of nitroglycerin (75–150 μg) increased the diameter of the left anterior descending coronary artery by 10%–20%, that of the left circumflex artery by 9%–22% and that of collaterals by 18%–28%, without significant changes in heart rate or blood pressure. A dose of 450 μg produced only a modest additional increase in coronary diameter but significantly increased heart rate and decreased blood pressure.

It should be remembered that coronary dilation must be selective in order to relieve rather than aggravate ischemia. It is accepted at present that dilation of the large capacitance coronary arteries is beneficial as these vessels are those affected by atherosclerosis. Moreover, ischemia does not by itself cause dilation of the capacitance vessels. In contrast, myocardial ischemia produces near-maximal dilation of coronary resistance vessels, mainly coronary arterioles, in the affected zone. Pharmacologic dilation of resistance coronary vessels can result in coronary steal from the ischemic zone to normal myocardial zones, which do not need an increase in blood supply. For this reason potent dilators of coronary arterioles such as dipyridamole are not used in angina pectoris. Dilation of large conduit coronary arteries may not contribute substantially to coronary flow at rest in the normal myocardium. However, when coronary arterioles are dilated due to ischemia, dilation of the large conduit coronary arteries will improve coronary flow. Dilation of the large coronary arteries, but not that of coronary arterioles, contributes to collateral flow to the ischemic zone and prevents coronary steal and impaired distribution of coronary flow to the deep layers, which may occur if peripheral coronary resistance decreases. According to these considerations nitrates have a near-optimal profile. Nitroglycerin dilates the normal epicardial and intramyocardial coronary arteries, coronary collaterals and, in the opinion of many investigators, also diseased, stenotic coronary arteries.

Angiographic studies have confirmed that normal coronary arteries are dilated by nitroglycerin. The findings of Feldman and coworkers were described above. Brown and coworkers [16] have shown an average increase of 18% in the cross-sectional area of normal coronary segments after sublingual administration of nitroglycerin.The increase was greater (35%) in large vessels, 1.6–2.3 mm in diameter, and only 9% in small vessels, 0.4–0.5 mm in diameter. Other investigators have also confirmed a greater dilatory effect of nitroglycerin on large coronary arteries than on small ones [17, 18]. Brown and coworkers [16] also demonstrated that nitroglycerin dilated stenotic lesions in large coronary arteries. An increase of 36% in cross-sectional area was observed in severe lesions and one of 22% in moderate lesions. The absolute increase, however, was greater in the moderate lesions.

Simon et al [19] have shown that nitrates dilate not only the coronary arterial bed. They studied the effect of 0.2 mg nitroglycerin administered directly into the left coronary artery in seven patients, three with and four without coronary disease. The transient drop in coronary resistance and subsequent increase in coronary flow after intracoronary nitroglycerin were associated with a marked increase in coronary vascular volume, which was still present when the effect on the coronary arteriolar bed had subsided. These findings imply a more widespread dilatory effect of nitroglycerin on the coronary circulation, with a different time course in different vascular segments.

Simonetti et al [19a] have observed a biphasic response of normal human coronary arteries to nitroglycerin. The drug was injected into the coronary arteries at doses of 50 and 300 mcg, 10 minutes apart. A marked dilatation of the microvessells, promptly followed by a mild decrease in flow velocity, and sustained dilation of the large conduit coronary vessels were observed. The investigators suggested the failure to recognize and determine the time of this response may explain conflicting results reported by previous studies.

Klein and colleagues [20] measured coronary flow through saphenous vein bypass grafts during surgery and evaluated changes in the native coronary circulation as reflected by flow in the grafts. They reported that intravenous nitroglycerin dilated diseased coronary arteries and coronary collaterals and improved perfusion of the ischemic myocardium.

Relief of coronary spasm by nitrates is, of course, also beneficial.

Two groups of investigators measured the effect of nitrates on blood velocity in the coronary arteries. In 1971 Benchimol et al [21] reported an increase in peak velocity of coronary flow over 3 min from sublingual administration of nitroglycerin. A more recent study by Simon et al [22] revealed that intracoronary and systemic administration of nitroglycerin decreased blood velocity in the coronary arteries. In coronary bypass grafts direct application of the drug increased blood velocity, and sublingual administration decreased it.

Nitroglycerin also improves perfusion in myocardial zones dependent on collateral flow, and in this effect it may be superior to other vasodilators. Fujita et al [23] evaluated the acute effects of nitroglycerin and diltiazem on collateral blood flow and regional myocardial function in the ischemic zone during a 2-min coronary occlusion in dogs. During the control of coronary occlusion ischemic segment shortening decreased to 33% ± 9% of preocclusion resting values. Following release of the occlusion, blood flow debt repayment was 166% ± 53%. Coronary blood flow during occlusion of the left anterior descending artery was at 20% ± 5% of the level of flow during resting in the left circumflex artery. These responses were not altered by diltiazem, 50 μg/kg intravenously. However, nitroglycerin, 5 μg/kg intravenously improved ischemic zone shortening to 58% ± 14%, decreased blood flow debt repayment to 74% ± 26% and increased coronary blood flow during occlusion to 30% ± 8%. Thus, nitroglycerin produced greater effects than diltiazem in increasing coronary blood flow and regional myocardial function in the collateral dependent zone when the effects of both drugs on the systemic circulation were minimized by the use of small doses.

Another method to approach this question is by myocardial perfusion studies. To resolve whether ISDN could enhance myocardial perfusion during exercise, Tono-oka et al [24] performed exercise Tl-201 myocardial scintigraphic studies and exercise radionuclide ventriculographies in 22 patients with severe coronary artery disease, before and after ISDN infusion, 0.1 mg/kg per hour. There was no significant difference in end-diastolic volume before and after ISDN, although the initial Tl-201 uptake index increased significantly from 77.6% ± 4.5% to 82.3% ± 4.9%, and the redistribution index decreased from 1.89 ± 1.12 to 0.45 ± 0.3. Thus, it may be concluded that ISDN actually enhances myocardial perfusion in patients with coronary artery disease during exercise.

Okazaki et al [24a] studied the effect of nitroglycerin on coronary collateral function in 21 patients with obstructive lesions in the left anterior descending coronary artery. Ten of them had collaterals and 11 did not. Two exercise tests were performed, before and after sublingual administration of nitroglycerin. Coronary blood flow was measured by the thermodilution method. Exercise duration increased by 90 ± 28% in the patients with collaterals and by 25 ± 19% only in those without collaterals. The corresponding increases in coronary blood flow were 41 ± 20% and 18 ± 12%, respectively. These findings indicate that nitrates improve exercise tolerance also by an enhanced coronary flow through collateral vessels.

Certain findings, however, do not accord with the theory that direct coronary dilation is the mechanism of antianginal effect of nitrates. Several investigators have reported that nitrates were unable to dilate the most severe coronary lesions (stenoses of 90% or more), which are most likely to produce angina pectoris [25, 20]. Moreover, Melchior et al [26] have recently shown that in patients with stable angina pectoris and an isolated proximal coronary stenosis, studied by cinedensitometry, sublingual nitroglycerin induced vasodilation in normal segments of the coronary tree but not in stenotic ones. This resulted in an increase in the degree of stenosis. The action of nitroglycerin in this group could, therefore, be explained essentially by its systemic effects (this does not necessarily apply, however, to patients with more complex patterns of coronary stenoses).

Other investigators have found that nitrates decreased coronary blood flow during exercise in patients with coronary artery disease [27]. It was also shown that intracoronary nitroglycerin failed to relieve pacing-induced angina, whereas intravenous nitroglycerin relieved the symptoms [28].

These findings indicate that systemic mechanisms may predominate in the antianginal effect of nitrates. This relates, however, to the types of angina which result from an increase in myocardial oxygen demand. In contrast, the action on the epicardial coronary vessels may play a major role in the management of angina at rest [29]. This is further discussed in the sections on unstable angina and coronary spasm.

In view of the new concept of mixed angina, it is no longer as simple to make the conventional distinction between the mechanisms of chronic stable angina pectoris and coronary spasm. Moreover, recently Krayenbühl et al [30] have shown that the narrowing of coronary artery stenosis during dynamic exercise may be attributed to active vasoconstriction. Evaluation by coronary angiography of 18 patients with stable angina pectoris demonstrated that this vasoconstriction is prevented by preexercise intracoronary nitroglycerin. It was also shown that coronary artery stenosis vasoconstriction can be an important mechanism of myocardial ischemia during dynamic exercise in patients with stable angina pectoris. These recent findings tend to balance to some extent the bulk of evidence emphasizing the predominance of systemic mechanisms in the antiischemic effect of nitrates.

These groups of investigators have reported later that diltiazem, injected into the coronary arteries, and nitroglicerin sublingually seem to have an additive effect on coronary vasodilation of stenotic coronary segments and in preventing exercise-induced vasoconstriction [30a]. This indicates that intracoronary nitroglycerin did not achieve maximal dilatation of stenotic coronary arteries. The exact mechanism of the additive effect of nitroglycerin is not clear but it has been suggested that diltiazem restores endothelial function of the diseased coronary vessel [30a]. It is possible that in the presence of this effect, nitroglycerin can be more effective.

4.2. Systemic Mechanism

Nitrates can also relieve angina by a decrease in myocardial oxygen demand, by preload and afterload reduction. In patients with coronary artery disease nitroglycerin was shown to decrease left ventricular end-diastolic volume, pressure, and wall tension, both at rest and during exercise, thereby reducing myocardial oxygen demand [31]. This effect also allows better perfusion of the subendocardial myocardium. Myocardial oxygen demand is further reduced by the decrease in systemic vascular resistance and systemic arterial pressure. This mechanism has two problems:

a) decrease in coronary perfusion pressure and
b) reflex tachycardia, which increases myocardial oxygen demand.

The overall hemodynamic effect is, however, beneficial and results in a decrease in myocardial oxygen demand.

Fuchs and co-workers [32] differentiated direct and systemic mechanisms and demonstrated the predominance of the latter. They showed relief of pacing-induced angina concomitantly with a decrease in coronary flow and mean aortic pressure in 10 of 11 patients treated with 400–800 μg of nitroglycerin sublingually or 200 μg intravenously. However, intravenous administration of 78 μg in seven patients produced a slight increase in coronary flow, no change in aortic pressure and no relief of angina. These findings indicate that the antianginal effect of nitroglycerin results from its cardiac unloading effect.

Engel et al [33] attempted to differentiate the effects of nitrates on the coronary arteries from the systemic effect under conditions of myocardial ischemia, defined as angina pectoris associated with ST-segment depression of 1 mm or more, in 29 patients with coronary artery disease. They used the Tl-201 scintigraphy and the precordial xenon-133 clearance technique. They found that the prevention of exercise-induced myocardial ischemia is associated with decrease in coronary blood flow. The main mechanism of the effect of nitrate is, therefore, a decrease in myorcardial oxygen requirements. Nitrates reduce flow to normal areas and do not alter flow to ischemic zones. They also make the flow pattern more homogeneous.

In summary, the bulk of evidence supports the theory that systemic effects are the main mechanism of the antianginal activity of nitrates in most forms of angina pectoris. In vasospastic angina the direct coronary effect plays a more important role.

4.3. *Effect of Nitrates on Myocardial Contractility in Patients with Coronary Artery Disease*

In patients with coronary artery disease abnormalities in left ventricular regional contractility are frequently found. Numerous studies have revealed that nitrates can reverse these abnormalities and occasionally even normalize contractility of ischemic segments.

McAnulty et al [34] reported that sublingual nitroglycerin diminished left ventricular wall motion abnormalities in patients with coronary artery disease. Several investigators confirmed that nitroglycerin improves the function of viable hypokinetic myocardial segments but not that of nonviable dyskinetic myocardial segments [35, 36, 37]. This response to nitroglycerin may be used to assess the myocardial functional reserve. Helfant et al [38] and Chesebro et al [39] reported that regional left ventricular contractile responses to sublingual administration of nitroglycerin accurately predicted improvement in regional function after coronary bypass surgery which restored blood supply to these segments.

Pepine and coworkers [40] studied patients with left ventricular dyskinetic segments in the absence of symptoms or electrocardiographic evidence of myocardial ischemia. Angiocardiography revealed that the presence of dyskinesia was significantly associated with the presence of coronary artery disease in a vessel supplying the segment. Nitroglycerin improved contractility in these segments, suggesting that the dyskinesia results from silent ischemia. Thus, silent ischemia can be reversed by nitroglycerin.

4.4. *Effect of Nitrates on Right Ventricular Hemodynamics in Coronary Artery Disease*

Only a few studies on the effect of nitrates on right ventricular hemodynamics in patients with coronary artery disease have been performed. Karsch et al [41] studied by cardiac catheterization ten patients with coronary artery disease. None of them had valvular defects, arterial hypertension or cardiomyopathy. Nitroglycerin, infused intravenously at a rate of 3 mg/h, significantly reduced the end-diastolic and systolic volume indices. Ejection fraction and mean diameter shortening velocity remained unchanged. Systolic and end-diastolic right ventricular pressure, mean pulmonary artery pressure and pulmonary capillary pressure were significantly reduced 10 min after initiation of the infusion. The dp/dt and heart rate were not significantly altered. Cardiac index and systemic vascular resistance decreased, while pulmonary arteriolar resistance remained unchanged. The pronounced decrease in pulmonary arterial pressure resulted, therefore, not from an effect on the pulmonary vascular resistance but from a decrease in cardiac output.

4.5. *Effect of Nitrates on Parameters of Diastolic Cardiac Function*

The diastolic function of the heart, and mainly of the left side of the heart has been recently the subject of many investigations. The two methods mainly used in the non-invasive studies are radionuclide angiography and Doppler echocardiography. The parameters of diastolic function measured by these techniques are peak filling rate, time to peak filling rate (in msec), time to peak filling rate expressed as percentage of diastole, the amount of left ventricular filling occurring during the first one third of diastole and others.

Employing these technique it was shown that in hypertension, ischemic heart disease, hypertrophic cardiomyopathy and concentric myocardial hypertrophy the diastolic function is impaired. The peak filling rate and first-one-third filling were found to be reduced, and the time to peak filling rate was prolonged.

Plotnick et al [41a] evaluated the effect of sublingual nitroglycerin on diastolic ventricular function in healthy human subjects. In the upright position nitroglycerin reduced the peak filling rate by 26 ± 5%. This effect was opposite to that produced by intravenous verapamil, despite similar changes in heart rate and blood pressure in these subjects. Nitroglycerin increased the ejection fraction and decreased the end-diastolic and end-systolic volumes [41a].

Nitrates affect also the right ventricular diastolic function. In patients with dilated cardiomyopathy or mitral value disease nitroglycerin was shown to reduce by 40% the stiffness constant of the diastolic pressure-volume relationship [41b].

4.5.1. Effects of Nitrates on Diastolic Properties of the Heart in Patients with Coronary Artery Disease

Nitrates were reported to change the diastolic function of the heart, as evidenced by a downward shift of the left ventricular diastolic pressure-volume curve [42].

Until recently however, it was not clear, whether this change is due to improved relaxation. In 1983 Amende et al [43] reported a study on the effect of intracoronary and sublingual nitroglycerin on left ventricular diastolic properties in patients with coronary artery disease. They found that sublingual nitroglycerin increases the rate of isovolumic relaxation and improves diastolic function, but these effects are not expressed by the negative dp/dt since this parameter depends on several hemodynamic factors. Intracoronary nitroglycerin accelerates the isovolumic relaxation but has no effect on diastolic function beyond the period of isovolumic relaxation.

Hirzel et al [44] reported that orally administered IS-5-MN increases the speed of left ventricular relaxation, as evidenced by a decrease in the time constant of left ventricular pressure decay. As the time constant is independent of preload and afterload, its decrease represents improvement of left ventricular relaxation by IS-5-MN, either directly or by relief of myocardial ischemia. These effects were achieved without any change in myocardial contractility. The faster pressure decrease produced by IS-5-MN is beneficial as it leads to prolongation of the effective diastolic coronary perfusion time.

4.5.2. The Effect of Nitrates on the Left Ventricular Diastolic Function in Hypertensive Patients

In the recent decade a remarkable research on diastolic properties of the left ventricle in hypertension has been performed. The most common abnormalities reported were reduction in peak filling rate in early diastole and prolongation of the time to peak filling as an index of left ventricular relaxation in early diastole [44a, 44b, 44c]. These abnormalities preceded abnormalities in systolic function.

Nitrates reduce mainly preload, and as such they may be used for the assessment of the relationship between changes in preload and diastolic function in hypertension. Furthermore, nitrates are used in combination with antihypertensive drugs in many patients with hypertension associated with ischemic heart disease. Therefore, it would be important to know the effect of nitrates alone on diastolic function in hypertensives and the interrelationship of changes in preload and diastolic function.

One of us studied the effect of a single dose of isosorbide-5-mononitrate on parameters of diastolic function in hypertensive subjects in order to evaluate the effect of changes in preload on these parameters. Eighteen patients with essential hypertension (13 males and 5 females, mean age 62 ± 10.2) were studied. The mean duration of the hpyertension was 11.2 years.

After a one week washout period the patients underwent radionuclide ventriculography. Immediately thereafter placebo was administered and a second radionuclide ventriculography was performed, for reproducibility purposes. One hour after the placebo administration 20 mg isosorbide mononitrate was given orally and a third radionuclide ventriculography performed.

Heart rate was not significantly altered. Mean blood pressure decreased significantly from 119 ± 17 to 101 ± 18. Left ventricular ejection fraction increased slightly but not significantly from 64.8 ± 9 to 68.3 ± 3.8%. Peak filling rate decreased from 15045 counts/sec to 13569 counts/sec ($p > 0.05$). Normalized peak filling rate was unaltered. Time to peak filling rate increased from 176 ± 36 to 195 ± 29 msec ($p > 0.01$) and when expressed as percent of diastole it significantly increased from 46.1 ± 11 to 53.7 ± 14% ($p > 0.01$). The filling rate of left atrium decreased from 7150 ± 1230 counts to 6287 ± 1264 counts. Left ventricular end-diastolic counts and stroke counts decreased from 4565 ± 765 to 4288 ± 901, respectively. There was no significant change in the right atrial filling rate or the end-diastolic and stroke counts of the right ventricle.

Thus IS-5-MN reduced slightly and nonsignificantly the normalized peak filling rate and significantly prolonged time to peak filling rate and percent of time to peak filling rate from total diastole. These results are similar to those reported by Yamagishi et al [44d] on the effect of sublingual ISDN on the diastolic function of normal persons.

It is likely that the smaller amount of blood returning to the left ventricle under lower pressure induced by the nitrate may account for the decrease in peak filling rate and the delay to peak filling observed in our study.

In conclusion, the study shows that nitrates produce a prolongation of time to peak diastolic filling of the left ventricle. This effect is mainly due to reduction in venous return.

4.6. *Effects of Nitrates on Platelet Function*

Although nitrates have been used for over 100 years in treatment of angina pectoris, their effects on platelet function have only recently been studied, i. e. since the role of platelets in various manifestations of ischemic heart disease was recognized. This research was stimulated by the finding that other antianginal agents, including beta-adrenoreceptor blockers, verapamil and nifedipine, have antiplatelet effects.

De Caterina et al [45] studied platelet-rich plasma from 22 healthy volunteers which was incubated with various concentrations of ISDN and with the vehicle alone for 5–10 min. Thereafter platelet aggregation was induced by adenosine 5′-diphosphate (ADP), adrenaline, collagen, arachidonic acid and thrombin, and thromboxane B_2 levels in the plasma were determined. A dose-dependent decrease in platelet aggregation and thromboxane B_2 concentration was found. It was concluded that such concentrations of ISDN, which were higher than those usually achieved in clinical practice, exert a direct in vitro antiplatelet effect.

For in vivo evaluation of this effect, ISDN was administered intravenously at an infusion rate of 4 mg/h to 11 patients with angina pectoris and at a rate of 30 mg/h to eight other patients. The lower infusion rate caused only slight changes in heart rate and arterial pressure but markedly decreased the platelet aggregation, induced by ADP and adrenaline, and the circulating platelet aggregates. The higher infusion rate caused greater changes in heart rate and blood pressure but smaller changes in platelet function. At the end of this infusion a rebound increase in circulating platelet aggregates was observed. Thromboxane levels were unaltered.

The in vivo study demonstrated that the antiplatelet effect is evident even at concentrations which did not cause significant hemodynamic changes. The antiplatelet effect of ISDN was more prominent when the drug was given in vivo than in vitro. This might be related to hepatic transformation of ISDN to its mononitrate metabolites, which show pharmacologic activity. Another explanation is the in vivo production of prostacyclin, a potent antiplatelet agent, by nitrates [46].

An alternative explanation to the finding that ISDN inhibits platelet activity at lower doses in vivo, while in vitro it fails to elicit platelet aggregation inhibition unless used at suprapharmacologic concentrations, has recently been suggested by Rolland et al [47]. They have demonstrated an endothelium-dependent ISDN activity, related to the antithrombogenic endothelial cell defense (ATECD). Modulation of ATECD may offer a new approach for antithrombotic therapy. Rolland et al [47] have also shown that endothelial cell dependent inhibition of platelet aggregation is increased more with ISDN than with IS-5-MN and is absent with isosorbide-2-mononitrate (IS-2-MN). It is important to note also that n-acetylcysteine, which has been shown to potentiate the hemodynamic effects of nitroglycerin and limit the development of hemodynamic tolerance, also potentiates the antiplatelet effects of this drug. The clinical relevance of the antiplatelet activity of nitrates in their mechanism of action in ischemic heart disease is yet to be determined.

Stamler et al [47a] found that platelets taken from patients treated with intravenous nitroglycerin manifest attenuated aggregation responses ex vivo, when thiol stores are repleted. These investigators proposed that the mechanism of action of nitroglycerin is mediated in part by its antiplatelet effect and that this effect depends on the adequacy of reduced intracellular thiol stores.

4.7. Effect of Nitrates on Blood Rheology

One of the responses to several vasodilators, including nitrates, is homodilution, due to shift of extravascular fluid into the intravascular compartment. Despite the reduction in hemoglobin concentration, associated with a decrease in viscosity, hemodilution can improve blood flow and oxygen supply. There is also evidence that hemoglobin values at the upper range of normal may be a cardiovascular risk factor and that the optimal values for the microcirculation is lower than the accepted normal values.

Brügger et al [47b] studied the effect of transdermal nitroglycerin, applied to 10 healthy volunteers for 10 days. The blood viscosity and hematocrit were significantly reduced. The investigators suggested that the decrease in viscosity induced by nitroglycerin resulted from internal hemodilution. The rheological improvement may contribute to the anti-ischemic properties of nitroglycerin.

5. Nitrates Compared and Combined with Other Antianginal Drugs

5.1. ISDN Compared with Other Antianginal Agents

As nitrates were introduced to clinical practice before all other classes of antianginal drugs, the development of new classes of drugs should have been followed by comparative trials with nitrates. The situation was complicated, however, by several factors. When beta-adrenoreceptor blockers were developed and introduced to clinical practice, organic nitrates were not in wide use for angina prophylaxis because of the mistaken belief that the first-pass hepatic degradation prevents orally administered organic nitrates from reaching the systemic circulation. Nitroglycerin was used only for relief of acute pain – an indication for which orally administered beta-blockers are ineffective. Long-acting nitroglycerin formulations and transdermal nitroglycerin patches were not available until 4 years ago. Therefore there are very few comparative studies of beta-blockers and nitrates in angina pectoris. Many comparative studies of nitrates and calcium antagonists have, however, been performed, most of these with ISDN.

5.1.1. Calcium Antagonists

Both nitrates and calcium antagonists act by systemic and central mechanisms. The systemic mechanisms include preload and afterload reduction. Afterload reduction is more pronounced with calcium antagonists and preload reduction with nitrates. The central mechanisms include coronary dilation and relief of coronary spasm. Calcium antagonists also have a negative inotropic effect which might further decrease the myocardial oxygen consumption.

Some calcium antagonists, such as nifedipine, can accelerate heart rate to the same extent as nitrates or to an even greater degree, due to their very potent vasodilatory properties. Other calcium antagonists, such as verapamil and diltiazem, do not accelerate heart rate due to less potent vasodilatory properties and/or direct suppressant effects on the sinoatrial node.

Clinical studies have revealed that in patients with exercise-induced angina pectoris ISDN and nifedipine as monotherapy exert comparable antianginal effects [48, 49]. The effect was comparable when both drugs were added to propranolol [50]. Schultz et al [51] suggested that the antianginal effect of ISDN and nifedipine may be related to the coronary anatomy. This is suggested by findings that nitrates dilate large coronary arteries and coronary stenoses, whereas calcium antagonists dilate both large and stenotic coronary arteries as well as arteriolar resistance vessels. Therefore calcium antagonists may cause coronary steal which might be deleterious in those patients with coronary artery disease who are prone to coronary steal. These investigators, in fact, called nitrates "benign vasodilators" and calcium antagonists "benign and malignant vasodilators." They found that in patients who are prone to coronary steal due to their coronary anatomy (coronary stenoses, collateral vessels and violable myocardium) ISDN was found to exert an antianginal effect whereas nifedipine did not.

It is possible that in some patients with angina pectoris who do not respond to nifedipine (about 15% of patients) the deleterious effect on coronary steal counteracts the beneficial effect of the drug. ISDN might be tried in these patients.

Engby et al [51a] reported that IS-5-MN, 60 mg slow release, and diltiazem 180 mg/day had a similar antianginal effect in 32 patients with chronic stable angina pectoris.

An important group of patients with angina pectoris are those with left ventricular dysfunction. Selection of antianginal agents for these patients depends not only on their efficacy but also on their effect upon ventricular function. Packer et al [52] compared the effects of single doses of nifedipine, 20 mg orally, and ISDN, 40 mg orally, in 16 patients with left ventricular dysfunction (ejection fraction less than 30%). Hemodynamic measurements were made at peak effects of each drug.

Nifedipine reduced systemic vascular resistance by 50% and ISDN by only 19%. However, the greater effect of nifedipine was translated into a greater decrease in mean arterial pressure and not a greater increase cardiac index or stroke volume index, which increased similarly with both drugs. ISDN, but not nifedipine, reduced left ventricular filling pressure and right atrial pressure. Hence, left ventricular performance improved with ISDN but not with nifedipine. The investigators attributed this difference to the negative inotropic effect of nifedipine, which prevented translation of the marked decrease in systemic vascular resistance into an increase in cardiac output.

The impairment of left ventricular function by nifedipine was associated with clinical deterioration in four patients. No patient experienced a deterioration while on ISDN. It was concluded that in patients with left ventricular dysfunction ISDN improves left ventricular performance to a greater degree and is better tolerated hemodynamically and clinically than nifedipine. Therefore ISDN should be preferred over nifedipine as first-line antianginal therapy in patients with left ventricular dysfunction.

Stegaru et al [53] studied the effects of long-term treatment with 120-mg sustained-release ISDN and 60-mg sustained-release nifedipine on myocardial perfusion in 40 patients with coronary artery disease, randomized to either treatment for 4 weeks. In both groups myocardial ischemia was reduced by therapy, but ISDN improved myocardial perfusion (evaluated by scintigraphy) to a considerably greater extent than did nifedipine.

5.1.2. Beta-Blockers

Beta-adrenoreceptor blockers and nitrates exert their antianginal effects by means of different mechanisms. Nitrates increase myocardial oxygen supply by coronary vasodilation and decrease myocardial oxygen demand by preload and afterload reduction. Beta-blockers have no coronary dilatory effect. They decrease myocardial oxygen demand by slowing heart rate and depressing myocardial contractility. Therefore, it is likely that some patients will respond to one drug but not to the other. Unfortunately, most comparative studies of nitrates and beta-blockers were performed in the early days of development of beta-blockers, using methods which are today not considered optimal. Since that time newly developed beta-blockers have been compared with propranolol but not with nitrates.

Conventionally, nitrates are considered "step 1" in antianginal therapy. Use of either group of drugs should depend on individual clinical evaluation, cardiovascular condition, safety profile, associated diseases, other drugs concomitantly used, etc.

5.2. ISDN Combined with Other Antianginal Agents

In patients who do not adequately respond to ISDN, combination with calcium antagonists and/or beta-blockers can be tried for prevention of angina. ISDN in the form of sublingual tablets or oral spray may be used for relief of pain in patients on long-term treatment with beta-blockers and/or calcium antagonists.

5.2.1. Calcium Antagonists

Until recently the combination of ISDN and calcium antagonists was used only for treatment of patients with coronary spasm and variant angina and in some patients with chronic stable angina pectoris in whom superimposition of coronary spasm is suspected. In recent years, however, this combination has frequently been used to treat chronic stable angina pectoris (although less frequently than the combination of ISDN and beta-blockers).

Hopf and coworkers [48] reported that the combination of low-dose (5–10 mg) ISDN with nifedipine enhanced the reduction in ST-segment depression induced by exercise. However, when higher doses of the nitrate were used, the addition of nifedipine produced further improvement in only a few patients and showed no effect in the group as a whole. It has been suggested that the addition of nifedipine was beneficial only in those patients with superimposed coronary spasm.

Severi et al [54] studied in an uncontrolled trial 120 patients with angina at rest and ST-segment elevation who were treated with verapamil and nitrates. Most patients showed symptomatic improvement which persisted for several years. About 50% of the patients were free of pain by the 4th year of treatment. Similarly to other forms of therapy this combined regimen failed to reduce the high incidence of acute myocardial infarction associated with this disease; 22% of patients developed acute myocardial infarction in the 1st year of treatment.

Raizner et al [55] reported that the addition of verapamil abolished anginal episodes in 15 of 17 patients with variant angina refractory to nitrates and produced marked improvement in the remaining two patients.

Lablanche et al [56] studied the additive effect of IS-5-MN, 20 mg, and nifedipine, 10 mg, in 12 patients with variant angina. All had ergometrine-induced coronary spasm during coronary arteriography. Within a period of 3 days they received either IS-5-MN or nifedipine and finally a combination of both agents.

An ergometric test with incremental doses was performed 60 min after drug administration. The combination of nifedipine and IS-5-MN was superior to nifedipine alone in all the six patients with a positive test and superior to IS-5-MN alone in all except one. These results support the hypothesis that nifedipine and nitrates may have additive antianginal effects and that their combination is superior to either of these drugs alone in the prevention of coronary arterial spasm.

Other investigators have not been able to confirm an additive antianginal efficacy of nitrates and calcium antagonists.

Abrams and Hoekenga [57] compared the effect of diltiazem monotherapy to diltiazem with oral nitroglycerin in a double-blind, placebo-controlled trial in 12 patients with chronic stable angina pectoris. The patients received diltiazem, 240 mg, diltiazem at the same dose plus oral nitroglycerin, 26–78 mg, and placebo for 2–3 weeks each. Exercise tests were performed 2.5 h after dosing. Addition of nitroglycerin did not enhance the anti-ischemic effect of diltiazem, as evidenced by the lack of further increase in time to onset of angina, time to 1-mm ST-segment depression or total duration of exercise. It was concluded that combination therapy with diltiazem and oral nitroglycerin offers no benefit to patients who respond to diltiazem alone.

It should be remembered, however, that the use of oral nitroglycerin is controversial, as the drug undergoes extensive first-pass hepatic metabolism. In the study by Abrams and Hoekenga [57] plasma concentrations of nitroglycerin were not measured. Hence, it is not known whether the high doses of nitroglycerin used were sufficient actually to overcome the first-pass hepatic degradation and reach the systemic circulation.

Engby et al [51a] found that the combination of IS-5-MN, 60 mg slow release, and diltiazem 60 mg 3 times daily did not offer an additional benefit over the use of either drug alone. This finding is supported by the report of Apollo et al [57a], who have recently shown that nifedipine neither attenuates nor accentuates the arteriolar dilatory or venodilatory response to nitrates.

In summary, the bulk of evidence indicates that the combination of nitrates and calcium antagonists may produce an anti-ischemic effect in patients with angina pectoris resistant to monotherapy with either of these agents. It is recommended to use nitrates as the first line of therapy, particularly in patients with left ventricular dysfunction, and to add calcium antagonists if the response is inadequate.

5.2.2. Beta-Blockers

As beta-blockers and nitrates act by different mechanisms in angina pectoris, their combination might increase the antianginal efficacy. Moreover, their combined use may eliminate problems associated with monotherapy. For example, sympathetic activation due to nitrate-induced vasodilation may prevent the excessive slowing of heart rate by beta-blockers. Coronary vasodilation by nitrates can counter the unopposed alpha-adrenergic stimulation in the coronary arteries of patients treated with nonselective beta-blockers.

Nitrates may be used with both selective and non-selective beta-blockers. It is not recommended to use them with beta-blockers possessing intrinsic sympathomimetic activity because of a potential risk of accelerating heart rate.

5.2.3. Calcium Antagonists and Beta-Blockers

The so-called "triple-regimen" of nitrates, calcium antagonists and beta-blockers is the maximal therapeutic regimen which can be used in ambulatory patients with chronic stable angina pectoris. This regimen combines the beneficial effects and different therapeutic mechanisms of the three classes of drugs, but at present it is not clear whether it is also an optimal regimen, as it also combines the problems inherent in the use of each of these agents.

Tolins et al [58] reported that although the addition of nifedipine or ISDN to propranolol was beneficial in patients with chronic stable angina pectoris, combined administration of all three drugs did not necessarily result in optimal therapy. Exercise duration was actually shortened with triple therapy compared to double therapy. Katz [59] reported that the addition of nifedipine, 60 mg/day, to combined therapy with propranolol and nitrates in maximally tolerated doses did not alter exercise tolerance, ST-segment depression, heart rate or blood pressure.

It is possible that the selection of patients has biased the results of studies with triple therapy, as those with very severe disease have been chosen for this therapy. Despite these results it is a common and often also an effective practice to use triple therapy in patients with angina pectoris who do not respond to a combination of two of the drugs.

Silber et al [60] demonstrated that exercise-induced myocardial ischemia in patients with insufficient response to nitrates can be significantly improved by verapamil as well as by propranolol. Since verapamil and propranolol were equally effective when added to ISDN, the reduction of contractility appears to be the prevailing mechanism for a successful anti-ischemic therapy in patients with insufficient response to nitrates. Silber et al [60] reported that these patients usually need a triple therapy for optimal anti-ischemic effect. This can be achieved without a significant deterioration in left ventricular ejection fraction during exercise.

5.2.4. Indoramin and Beta-Blockers

Indoramin is a selective alpha-1 adrenoreceptor blocker, used mainly for treatment of hypertension. Collins and Sheridan [61] studied the effect of indoramin in 15 patients with chronic stable angina pectoris. Thirteen of them had been receiving nitrates and beta-blockers, and this treatment was continued unchanged throughout the study. At a dose of 25 mg three times daily indoramin prolonged exercise duration and increased oxygen consumption during exercise, while the maximal double product was unchanged. The increased exercise capacity was associated with a reduction in ST-segment depression during exercise. These results indicate that indoramin may provide useful additional benefit to patients with stable angina pectoris who are already receiving nitrates and beta-blockers.

5.3. Nitroglycerin Compared with Other Antianginal Drugs

5.3.1. Calcium Antagonists

Both sublingual nitroglycerin (or oral spray) and sublingual and chewable nifedipine are used for rapid relief of pain in chronic stable and vasospastic angina pectoris. Current practice in chronic stable angina is to first use sublingual nitroglycerin and, if ineffective, to give nifedipine. In vasospastic angina it is still controversial as to which drug should be used first.

Angiographic studies have revealed that nifedipine, 20 mg, or nitroglycerin, 0.8 mg, administered sublingually, similarly dilated coronary stenoses [62]. The combination of both drugs dilated coronary stenoses more than either drug alone. Each drug dilated coronary stenoses by about 30% in approximately one-half of the obstructions. The remaining one-half of stenoses did not respond to either drug. Both drugs given together produced a dilation of about 50% in more than half the stenoses. Responding and nonresponding lesions were found in the same patients. Nitroglycerin has the advantage of a more rapid onset of action than nifedipine. However, the antianginal effect of nifedipine lasts longer than that of nitroglycerin.

In a recent study nitroglycerin was found superior to diltiazem [63] in patients with chronic stable angina pectoris. Intravenously administered nitroglycerin, but not diltiazem, relieved exercise-induced angina and reversed ischemic electrocardiographic changes. In another study conventional sublingual doses of nitroglycerin exerted an antianginal effect similar to that of an oral dose of diltiazem, 90 mg.

Sublingual nitroglycerin is definitely preferred over calcium antagonists for relief of acute pain in chronic stable angina pectoris, mainly because of the more rapid onset of action. If the response is not sufficient, sublingual nifedipine may be used. It is not known whether nitroglycerin (in the form of transdermal patches) or calcium antagonists are more effective for angina prophylaxis. Further studies are required to clarify this point.

Recently Choong et al [63a] reported a study of the antianginal effect of nitroglycerin and nifedipine, utilizing measurements of systemic and right heart pressures, oxygen consumption, exercise tolerance and radionuclide left ventricular ejection fraction and volume in 14 patients with chronic stable angina pectoris. Exercise tests were performed before and after administration of nitroglycerin intravenously and nifedipine orally, at doses which lowered mean arterial pressure by 20 mmHg. Exercise tolerance improved significantly more with nitroglycerin than with nifedipine. At submaximal workload heart rate was higher on nifedipine. At peak exercise on nitroglycerin heart rate, rate-pressure product, cardiac index and oxygen consumption were significantly increased over peak values on nifedipine or without therapy. Nitroglycerin reduced the pulmonary capillary wedge pressure more and systemic diastolic pressure less than nifedipine. Thus the coronary perfusion gradient was maintained by nitroglycerin and reduced by nifedipine. Nitroglycerin was associated with less angina and ST-segment depression than nifedipine or no treatment. The left ventricular diastolic pressure-volume relationship was improved by nitroglycerin but not by nifedipine.

These results do not necessarily indicate that nitrates are better antianginal agents than nifedipine. The fact that at comparable lowering of mean arterial pressure the anti-ischemic effects of nitroglycerin were better than those of nifedipine suggests that the anti-ischemic effect of nitroglycerin is not only attributed to reduction of myocardial oxygen demand but results also from other mechanisms such as improved oxygen supply.

5.3.2. Beta-Blockers

Sublingual nitroglycerin cannot be compared with beta-blockers because they are not used for the same indications. A comparative study of intravenously administered nitroglycerin and esmolol in unstable angina is now planned. Comparative studies of transdermal nitroglycerin patches and beta-blockers in long-term prevention of angina have not been performed.

5.4. General Considerations in Comparative Evaluation of Antianginal Therapy

The availability of three different classes of antianginal agents (nitrates, beta-blockers and calcium antagonists), each with distinct pharmacologic properties, provides the cardiologist with the opportunity to use a selective and rational approach to the individualized therapy of angina pectoris. The development of new and sophisticated drug delivery systems provides the opportunity to use these drugs in convenient and simple therapeutic regimens.

The selection of specific drugs for patients with angina pectoris depends on the mechanism of ischemia, the presence of associated cardiovascular problems and/or other diseases, the specific indications for which the drug is used, contraindications to the drug, tolerance to the drug, simplicity of therapeutic regimens, and cost of the drug. In the specific case of antianginal drugs associated therapeutic goals, such as the role of the drug in preventing infarction and sudden death and in treating hypertension, should be considered.

5.5. Mechanism of Angina and Selection of Drugs

In patients with angina at rest the main mechanism of ischemia is vasospasm in the coronary arteries. As vasospastic angina was extensively investigated at the time of introduction of the calcium antagonists, particularly nifedipine, into clinical practice, these agents became associated, in the opinion of many clinicians, with treatment of vasospastic angina. However, in vasospastic angina nitrates are at least equally effective to calcium antagonists, and occasionally even more effective. Some patients respond better to nitrates and others to calcium antagonists [64, 65]. Although beta-blockers might be effective in some patients with vasospastic angina, there are theoretical considerations not to use beta-blockers in this condition. The unopposed alpha-adrenergic stimulation in the coronary arteries in these cases might stimulate or enhance coronary spasm. In unstable angina with a mixed mechanism of fixed coronary stenosis and coronary spasm beta-blockers may be superior to nifedipine [66]. However, the theoretical considerations for not using beta-blockers are still valid. Therefore it is preferable to initiate therapy with nitrates.

It is not easy to perform accurate comparative studies of antianginal drugs in angina pectoris on effort resulting from fixed coronary stenosis. Most comparative studies have been single-dose or short-term studies, whereas treatment of stable angina pectoris is chronic. In most studies inadequate dose-titrations have been performed. This was the case particularly in many of the recent studies comparing nitrates to other antianginal drugs. It has recently been shown that the antianginal efficacy of nitrates is dose-dependent [67]. In many comparative studies suboptimal doses of ISDN have been used. Moreover, many of the comparative studies were performed in patients who did not respond to one class of drugs and not in newly treated patients. All these methodological problems complicate the comparative evaluation of nitrates, beta-blockers and calcium antagonists in chronic stable angina pectoris.

In a recent review on comparisons and combinations in antianginal therapy Julian [68] stated that nitrates, beta-blockers and calcium antagonists are about equally effective in their ability to prevent exercise-induced angina. However, he emphasized an important problem of nifedipine, stating that "nifedipine is probably equally effective provided that the optimal dose for the particular individual is chosen." The optimal dose of nifedipine shows interpatient variability, and accurate determination of the optimal dose might be crucial with this drug. Increasing the dose of nifedipine might decrease rather than increase the antianginal effect [69].

These conclusions are supported by several comparative studies of nitrates and other antianginal agents, described earlier in this chapter.

5.6. Associated Impairments in Cardiovascular Function

The advantages of nitrates are further emphasized when associated impairments of cardiovascular function are considered. In patients with angina pectoris associated with heart failure or impairment in left ventricular function beta-blockers are contraindicated. Although calcium antagonists may improve myocardial performance in some of these patients, it is not rational to use agents with a negative inotropic effect when agents with no such effect may be used. Moreover, nitrates are an established therapy for heart failure and in patients with combined angina pectoris, and heart failure monotherapy with nitrates is effective for both conditions. It should also be remembered that in some cases nifedipine has actually aggravated heart failure.

At present there is one theoretical advantage for the use of beta-blockers in patients with postinfarction angina - the possibility of secondary prevention. Propranolol, timolol and metoprolol have been found effective for this indication. There are, however, three problems with this approach:

1. It is not known whether beta-blockers exert their protective effects for periods longer than those studied in the secondary prevention trials.
2. About one-third of the patients, or even more, can not receive beta-blockers for this indication.
3. In patients with transient heart failure during acute myocardial infarction beta-blockers may provoke recurrences of heart failure late after the acute event.

In patients with Raynaud's phenomenon or intermittent claudications beta-blockers should not be used. In patients with pulmonary obstructive diseases nitrates and calcium antagonists do not impair and might even improve pulmonary function. Nonselective beta-blockers should not be used in such patients. Patients with bradycardia or conduction disturbances should, preferably, receive nitrates or nifedipine rather than beta-blockers, verapamil or diltiazem. The therapeutic regimen in patients with hypertension should, preferably, include beta-blockers or calcium antagonists.

5.6.1. Tolerance to Antianginal Effect

Tolerance to various effects of nitrates is discussed in other chapters of this book.

5.7. Individualized Antianginal Therapy Guided by Pretreatment Systemic Arterial Pressure

5.7.1. Overlap of Antianginal and Antihypertensive Treatment

All drugs used for treatment of angina pectoris are effective also in antihypertensive treatment. This goes without saying for drugs such as beta-adrenoreceptor blockers and calcium antagonists, which are considered first line antihypertensive drugs. Even nitrates are effective in certain forms of hypertension, such as isolated systolic hypertension in the elderly. Drugs that have been developed purely for treatment of hypertension, such as the angiotensin converting enzyme inhibitors (ACEI) are now evaluated for treatment of angina pectoris, with variable results. Even alpha-adrenoreceptor blockers such as prazosin were found effective in certain forms of angina pectoris.

Antianginal drugs act mainly by two mechanisms:

1. Reduction in myocardial oxygen demand;
2. Increase in myocardial blood supply.

It is the delicate interplay between the pathophysiology in any individual patient and the relative effect of drugs on either mechanism that determines the suitability of a specific drug for an individual patient. Pretreatment systemic arterial pressure is a determinant of myocardial oxygen demand (afterload) and supply (coronary perfusion pressure). Obviously it is a major factor affecting the therapeutic response to various antianginal drugs.

Except for nitrates all antianginal drugs are also antihypertensive drugs. Calcium antagonists and beta-adrenoreceptor blockers are even first-line antihypertensive drugs. Although their blood-pressure lowering effect is proportional to pretreatment blood pressure, they can lower blood pressure also in normotensives. Nitrates lower myocardial oxygen demand by both afterload and preload reduction, and increase coronary supply by coronary dilation. Although they can lower blood pressure they are not an antihypertensive drug.

5.7.2. Calcium Antagonists

The main mechanism of the antianginal effect of calcium antagonists is afterload reduction and lowering of myocardial oxygen demand and not an increase in coronary blood supply. This statement is based on the finding that despite an increase in exercise duration calcium antagonists (nifedipine as well as diltiazem) do not increase the heart rate-systolic pressure product during maximum exercise [70a].

Blood pressure and myocardial contractility can be reduced only to a limited extent before the beneficial effect on myocardial oxygen demand is offset by a reduction in coronary perfusion and impairment of cardiac function. Therefore additional mechanisms, including coronary vasodilation and preload reduction, are important. In normotensive patients it is preferable to use antianginal agents which increase exercise duration without excessive reduction in the double-product and, mainly, in its blood pressure component.

The potential problems associated with the use of potent systemic vasodilators such as calcium antagonists in patients with coronary artery disease are emphasized by two findings:

1. The phenomenon of "paradoxical angina" – an increase in extent of angina in some patients with coronary artery disease treated with calcium antagonists for this indication [70b, 70c]. In one study nifedipine increased angina

in 6 of 24 patients with chronic stable angina pectoris [70d]. Although the real incidence of paradoxical angina is much lower, it is a serious adverse effect. Its most likely explanation is excessive lowering of blood pressure and coronary perfusion pressure;

2. The failure to show a consistent beneficial effect of calcium antagonists in secondary prevention [70e], mainly when given early in the course of acute myocardial infarction, is probably related to excessive lowering of blood pressure in some patients, mainly those who were normotensive before drug administration. This hypothesis is supported by the tendency to excess mortality in infarct patients with systolic pressure of 100 mmHg or below, suggested by the nifedipine secondary prevention trial [70f].

The significance of adverse effects associated with blood pressure lowering is emphasized when combined therapy with two calcium antagonists is compared to monotherapy with one such agent. In a study with nifedipine and diltiazem [70g] patients receiving combination therapy suffered more side effects than those on either nifedipine or diltiazem monotherapy. Most side effects including dizziness, palpitation, and aggravation of angina might have been related to lowering of blood pressure.

These findings indicate that calcium antagonists should not be the first antianginal drug in normotensive patients and certainly not in those who are hypotensive. Several clinical trials are now ongoing to compare the antianginal effect of dihydrophyridine calcium antagonists in normotensive and hypertensive patients. A trial comparing the antianginal effect of nitrates and calcium antagonists in normotensive and hypertensive patients is indicated.

5.7.3. Beta-Adrenoreceptor Blockers

Beta-adrenoreceptor blockers are first line antihypertensive agents. They have also a potent anti-anginal effect attributed to lowering of myocardial oxygen demand due to slowing of heart rate, decrease of contractility, lowering of blood pressure and other mechanisms of minor importance [70h, 70i, 70g, 70k].

Beta-adrenoreceptor blockers are excellent antianginal drugs in hypertensive patients and in the majority of normotensive patients. However, any agent that acts largely by lowering of blood pressure and depression of myocardial contractility must theoretically have limited efficacy in normotensive or hypotensive patients. This problem was not addressed in most studies evaluating beta-blockers in angina pectoris, as hypotensive patients were excluded and the results were not analyzed in relation to pretreatment blood pressure. It is possible that such an analysis would have revealed a greater effect in those with higher pretreatment pressure and a lesser, or even a negative effect, in normotensives and hypotensives.

This potential problem is seen also in the secondary prevention trials with beta-adrenoreceptor blockers [70l, 70m, 70o]. Overt and even borderline hypotensive patients have been excluded from these trial. Their inclusion might have, theoretically, caused some cases of death due to aggravation of ischemia, that could offset the beneficial effect of beta-blockers in other post infarction patients.

Studies comparing the antianginal effect of beta-blockers in hypertensive and normotensive patients are indicated. Until the results of such studies are available it is advisable to initiate treatment of angina pectoris in normotensive and, mainly, in hypotensive patients with other drugs.

5.7.4. Angiotensin Converting Enzyme Inhibitors

Animal experiments revealed a reduction of ischemic insults by angiotensin converting enzyme (ACE) inhibitors [70p, 70q, 70r, 70s]. The mechanism of this effect could be afterload and preload reduction, an inhibitory effect on local catecholamine release, an effect on tissue renin-angiotensin system and even the negative inotropic effect of these agents. In post myocardial infarction patients remodeling of the left ventricle can be modified by captopril [70t].

These finding have little relevance to treatment of angina pectoris. Any antianginal effect that ACE inhibitors might have is probably related mainly to afterload reduction. ACE inhibitors are first-line anti-hypertensive agents. The animal studies mentioned above did not take into account the relation between pretreatment blood pressure and the effect of ACE inhibitors on ischemia. This question should be addressed directly by clinical trials in normotensive and hypertensive patients with angina pectoris and other ischemic presentations.

The effect of ACE inhibitors on myocardial ischemia in normotensive patients has been studied only recently. Gibbs et al [70u] compared the effects of enalapril and placebo, each given for 2 weeks, in 12 such patients. Four of the patients prolonged their total exercise time by 20% or more. Two patients, however, developed ischemia earlier on exercise and reduce their total exercise duration. Their exercise blood pressure did not differ from the patients who improved. In conclusion, the effect of enalapril on myocardial ischemia is variable. Initiation of treatment with enalapril in normotensive patients with angina pectoris requires careful monitoring and at present could not, of course, be recommended clinically.

A more practical issue is the use of ACE inhibitors in normotensive patients with heart failure of ischemic etiology, with or without angina pectoris which are the largest group of heart failure patients. Gibbs et al [70u] stated that "enalapril should be used with caution in treating heart failure caused by ischemic heart disease, since the beneficial effects of ACE inhibition on the heart failure may be offset if the myocardial ischemia is made worse".

5.7.5. Nitrates

Nitrates are first line antianginal but not antihypertensive drugs. Their mechanism of action includes coronary vasodilation [70v, 19], preload and afterload reduction [32] and effects on the diastolic properties of the heart and on platelet function [45]. Nitrates do not rely so much on afterload reduction and lowering of blood pressure as do other antianginal agents. The effect on the coronary arteries is a "positive" vasodilation, i.e. dilation of the large coronary arteries including stenotic segments, thereby increasing the myocardial blood supply (in contrast to "negative" vasodilation of coronary resistance vessels, resulting in coronary steal). Nitrates also improve perfusion in myocardial zones dependent on collateral flow and in this effect they may be superior to other vasodilators, including calcium antagonists [70w]. They can prevent the exercise-induced coronary vasoconstriction observed in patients with coronary artery disease [30]. Nitrates also make the coronary flow more homogenous [33], thus favoring ischemic zones over normal zones.

The main differences between nitrates and calcium antagonists are the greater preload reduction (relatively to afterload reduction), the lack of a negative inotropic effect of nitrates and a somewhat different profile of the coronary effect.

Nitrates are not antihypertensive drugs. Although they lower blood pressure acutely, this effect is largely attenuated during chronic use. The only case in which nitrates are used, investigationally, for treatment of hypertension, is isolated systolic hypertension in the elderly [210]. In this condition the pathophysiology is different from that of the usual cases of essential hypertension, and is mainly related to stiffness of the conduit arteries. Even in this condition I was unable to find a marked blood pressure lowering effect of nitrates (unpublished) data).

All these findings indicate that nitrates are superior to other antianginal and anti-ischemic drugs for treatment of myocardial ischemia in normotensive patients. This recommendation is supported by the extremely low incidence of aggravation of ischemia during chronic oral nitrate therapy.

5.7.6. Combined Therapy

Combined therapy with beta-adrenoreceptors and calcium antagonists produces an additive antianginal effect only in patients with normal or mildly impaired left ventricular function who remain symptomatic despite optimal treatment with a single agent alone [70x]. It is unlikely that normotensive agents, mainly with impaired left ventricular function, would benefit from combined therapy with potent blood-pressure lowering drugs. Packer [70x] recommended that if patients with left ventricular dysfunction continue to experience pain during therapy with nitrates and beta-blockers, coronary artery bypass surgery rather than addition of a third drug should be considered. This is definitely the case for those of them who are also normotensive or hypotensive.

5.8. Discontinuation of Drugs

Withdrawal phenomena related to myocardial ischemia have been observed with beta-blockers and calcium antagonists. Whether problems exist with nitrate withdrawal has not been clearly established for angina.

5.9. Adverse Effects

Hypotension is a major adverse effect of all three classes of antianginal drugs. For calcium antagonists and beta-blockers this effect is usually sustained unless the dose is reduced. For nitrates, however, the initial hypotensive effect is almost always very rapidly attenuated.

Beta-blockers have specific adverse effects, such as cold extremities, vivid dreams and impotence, which are almost never observed with nitrates. Verapamil causes constipation which might be intolerable, particularly in elderly patients. Nifedipine causes edema in about 20% of the patients. Headache, which often accompanies the initiation of nitrate therapy might be intolerable in some patients. No large-scale studies have specifically compared the frequency of adverse effects of the three classes of antianginal agents.

5.10. Nitrates and Diagnostic Procedures in Patients with Coronary Artery Disease

The anti-ischemic effect of nitrates may increase the sensitivity of radionuclide studies in detecting viable myocardium in patients with coronary artery disease. For example, T1201 SPECT during ISDN infusion was shown to detect viable myocardium in an infarcted zone, indicating the feasibility of coronary artery bypass surgery or angioplasty [70y]. Good results were observed after revascularization in patients with a positive ISDN test.

6. Unstable Angina Pectoris

The introduction of intravenously administered nitrate formulations such as intravenous nitroglycerin and intravenous isosorbide dinitrate has markedly improved the conservative therapy of unstable angina. These formulations, originally used only in coronary care units, are now used in general medical wards and emergencies rooms, without hemodynamic monitoring, and in some countries also in mobile coronary care units.

The experience of my colleagues and myself has been mainly with ISDN. Patients with unstable angina refractory to conventional therapy often respond to intravenous ISDN; some patients respond even to a low infusion rate, 1–2 mg/h, despite refractoriness to high doses of oral nitrates (it is not clear whether pharmacokinetic considerations are the only cause of this differential efficacy). Other patients require infusion rates as high as 15 mg/h, which are the highest doses recommended by most investigators. In some centers, however, doses as high as 50 mg/h and more are used in resistant cases of unstable angina. The drug also suppresses the electrocardiographic changes of ischemia and reduces the consumption of morphine and other analgesics.

The effect of this treatment on the long-term course of unstable angina is not known. It is, however, no longer relevant in the era of interventional cardiology. Short-term stabilization of patients on ISDN is now usually followed by early catheterization, percutaneous transluminal coronary angioplasty or coronary bypass surgery.

Distante et al [71] studied the effect of intravenous ISDN during the hot phase of unstable angina in 15 patients admitted to a coronary care unit with angina at rest (at least four ischemic episodes per day) which was poorly controlled by oral and/or transdermal nitrates, calcium antagonists and beta-adrenoreceptor blockers. Following ISDN infusion (mean dose 3.7 mg/h) episodes of myocardial ischemia were completely abolished in 11 of the 15 patients. Mean duration of therapy was 8.3 days. After discharge 7 of the 11 responders remained asymptomatic.

Other investigators have more experience with intravenous nitroglycerin. Page et al [72] studied the effect of intravenous nitroglycerin in 67 patients with severe unstable angina, treated for 2–29 days with titrated doses. One-third of the patients had Prinzmetal's variant angina. Clinical condition improved in 95% of the patients and complete suppression of symptoms was observed in 63%. Intravenous nitroglycerin was used alone in 25% of the patients and together with beta-blockers in 50% of the patients. The antianginal effect was associated with a decrease of 14 mmHg in systolic arterial pressure and an increase of 3 beats/min in heart rate. Diastolic arterial pressure was not altered. This is a most important safety factor in patients with coronary artery disease, in whom reduction in coronary arterial pressure is deleterious.

Rafflenbeul et al [73] compared the effects of nifedipine, administered intravenously at a rate of 15–30 mg/24 h, and nitroglycerin, administered intravenously at a rate of 30–60 mg/24 h, over a 72-h observation period in 20 patients with angina pectoris at rest. In the ten patients treated with nifedipine systolic and diastolic blood pressure fell significantly from $149 \pm 10/95 \pm 7$ to $134 \pm 13/88 \pm 5$ mmHg, cardiac index rose from 3.0 to 3.6 L min^{-1} m^{-2}, and heart rate and pulmonary capillary pressure remained unchanged. These effects were sustained throughout the period of study. In contrast, in the ten patients treated with nitroglycerin alone a transient decrease in arterial pressure and a reflectory rise in heart rate were observed, while pulmonary capillary pressure showed a sustained decrease from 14 ± 2 to 8 ± 3 mmHg, and cardiac index was unchanged. Anginal symptoms were alleviated in both groups 15–30 min after initiation of the infusion and ceased within 2 h.

These findings indicate that while both intravenous nifedipine and nitroglycerin are effective in unstable angina, their mechanisms of action differ. Nifedipine acts mainly by afterload reduction and nitroglycerin by preload reduction. It should be remembered that excessive afterload reduction is associated with a decrease in coronary perfusion pressure, which may be deleterious in patients with unstable angina. These differences apply also to ISDN. IS-5-MN may be even more effective if the preliminary findings that it decreases systemic vascular resistance less than other nitrates are confirmed.

If nitrate monotherapy does not completely abolish symptoms and/or signs of ischemia in unstable angina, other antianginal agents should be added. The combination with calcium antagonists is very effective. In cases of unstable angina not related to coronary spasm beta-blockers may also be added. Their use in vasospastic angina is controversial, as it has been suggested that they may aggravate the spasm (due to unopposed coronary alpha-adrenergic stimulation).

N-acetylcysteine has been shown to potentiate the hemodynamic and antiplatelet effects of nitroglycerin and to limit the development of hemodynamic tolerance.

Horowitz et al [74] have shown that concomitant administration of intravenous nitroglycerin and N-acetylcysteine in patients with unstable angina pectoris may facilitate management both by limiting the doses of nitroglycerin required and by reducing the incidence of acute myocardial infarction. The effect on development of myocardial infarction may reflect the antiplatelet actions of this combination. Horowitz et al [74], however, have found a high incidence of severe hypotension with the combination of nitroglycerin and N-acetylcysteine, suggesting that this regimen should be used with caution.

The practice of continuous administration of nitrates has been questioned in patients with unstable angina because of the possibility of tolerance development. However, the beneficial effects of intravenous nitrates in unstable angina may be related more to direct coronary effects and/or to antiplatelet effects. In this clinical condition tolerance does not seem to be a problem and continuous therapy, even with high doses, is still recommended [74a].

In summary, nitrates are the most important agents available for intravenous administration in patients with unstable angina pectoris. The ability to terminate the effect within a short period after termination of the infusion is crucial in the rapidly changing hemodynamic conditions which might accompany unstable angina. Only recently have agents of other classes (such as the ultrashort-acting beta-blocker esmolol) been developed to achieve the same flexibility as is available with nitrates in rapidly changing conditions such as unstable angina pectoris.

6.1. Effect on Myocardial Metabolism in Unstable Angina

Another effect which may play a role (although a limited one) in the anti-ischemic activity of nitrates is the effect on myocardial metabolism. In order to characterize regional myocardial perfusion and glucose metabolism in unstable angina and the effect of ISDN on these parameters Araujo et al [75] studied 22 patients with angiographically proven coronary artery disease and severe unstable angina, using positron-emission tomography. Myocardial glucose utilization was found to be increased in these patients, even in the absence of acute myocardial necrosis, clinical signs of acute ischemia or detectable perfusion abnormalities. Intravenous administration of ISDN reduced myocardial oxygen utilization, suggesting a beneficial effect on myocardial metabolism.

6.2. Coronary Spasm

Until about 20 years ago angina was generally believed to result almost exclusively from increased myocardial oxygen demand in the presence of a fixed atherosclerotic coronary stenosis. Even after the first descriptions of angina at rest and of variant angina [76, 77] these conditions were considered to be limited to a small number of patients. Only in the past 15 years have they been recognized as affecting large numbers of patients.

Extensive studies of angina at rest followed the development of calcium antagonists, which were considered very effective in this condition. It was found that angina at rest and variant angina usually result from changes in coronary tone. The most important of these is coronary spasm: a local increase in coronary tone in a short segment of the vessel. However, also a more generalized increase in coronary tone, although of less extent than coronary spasm, can sometimes result in angina.

Recently nitrates have been shown to be very effective in relief of coronary spasm. Many patients with coronary spasm respond completely to a conventional sublingual dose of nitroglycerin [78, 79]. In certain patients sublingual nitroglycerin had only a partial or no beneficial effect. Some of these patients responded better to intravenous nitroglycerin [80, 81]. Transdermal administration was also found to be effective. For example, nitroglycerin ointment was shown to reduce the daily number of ischemic episodes from 12.5 ± 3.9 to 0.5 ± 0.4 in ten patients with vasospastic angina pectoris. The most effective form of nitrate administration is by the intracoronary route. Buxton and coworkers [82] reported that intracoronary administration of nitroglycerin relieved ergonovine-induced coronary spasm refractory to other forms of administration of the drug. Pepine and colleagues [83] reported that intracoronary nitroglycerin was effective in treating patients with spontaneously occurring coronary spasm refractory to sublingual and intravenous nitroglycerin. This effect was demonstrated in both angiographically normal as well as atherosclerotic stenotic coronary arteries.

Several large clinical studies have confirmed the efficacy of nitrates in angina at rest. For example, Curfman and coworkers [84] studied the effect of intravenous nitroglycerin, 10–200 μg/min given over 72 h, in 40 patients with angina pectoris at rest. The drug reduced the frequency of anginal attacks from 3.3 $\pm$ 0.8 to 1.0 $\pm$ 0.3 per 24 h. However, only 36% of patients showed complete suppression of angina, whereas 43% of patients were refractory to treatment and required early coronary artery bypass surgery. An encouraging finding for ambulatory treatment was that the effect of intravenous nitroglycerin was similar to that of combined treatment with oral ISDN and nitroglycerin ointment.

Kaplan et al [85] studied the effect of intravenous nitroglycerin in 35 patients with angina at rest refractory to oral and sublingual nitrates and beta-blockers. Nitroglycerin, 140 $\pm$ 15 μg/min, reduced the number of anginal episodes from 3.5 $\pm$ 0.4 to 0.3 $\pm$ 0.1 per day. Sublingual nitroglycerin consumption decreased from 1.9 $\pm$ 0.3 to 0.4 $\pm$ 0.1 mg/day. In 25 of the 35 patients complete suppression of anginal episodes was achieved while in eight patients the response was partial, and two patients did not respond.

ISDN is also effective for relief of coronary spasm. Until 5 years ago it was the only formulation available for continued treatment without using the intravenous route. Distante et al [86] reported that ISDN significantly reduced the frequency of ischemic attacks in patients with vasospastic angina pectoris. My personal experience includes several patients with variant angina due to coronary spasm who responded to monotherapy with intravenous ISDN.

Curfman et al [84] studied the effect of combined administration of oral ISDN, 20–60 mg four times daily, and nitroglycerin ointment in patients with rest angina. The number of anginal episodes was reduced from 3.1 $\pm$ 0.4 to 1.4 $\pm$ 0.3 per day. In 17% of patients the angina was completely abolished, however, 61% required early coronary bypass surgery.

Winniford et al [87] studied the efficacy of concomitant calcium antagonists and ISDN therapy in patients with frequent episodes of variant angina and compared this combination with ISDN monotherapy. Nine patients were treated with:

1. ISDN alone, 117 $\pm$ 63 mg/day
2. verapamil, 453 $\pm$ 75 mg/day, plus ISDN 117 $\pm$ 63 mg/day
3. nifedipine, 71 $\pm$ 14 mg/day plus ISDN 117 $\pm$ 63 mg.

Each regimen was given for 2 months. During ISDN therapy these nine patients had an average of 23.7 $\pm$ 37.3 episodes of chest pain per week, consumed 24.4. $\pm$ 47.4 sublingual nitroglycerin tablets per week and had 46.5 $\pm$

43.2 episodes per week of transient ST-segment deviations. During treatment with verapamil and ISDN the frequency of angina was reduced to 3.9 ± 3.6 episodes per week and that of ST-segment deviations to 3.5 ± 2.6 deviations per week. During treatment with nifedipine and ISDN the frequency of angina was 3.1 ± 4.0 episodes per week and ST-segment deviations to 5.5 ± 6.6 deviations per week. The effects of both combined regimens were similar to one another.

It may be concluded that in patients with severe variant angina and very frequent anginal episodes the combination of a calcium antagonist and ISDN is more effective than ISDN alone in reducing the frequency of anginal episodes and ischemic electrocardiographic alterations. It should be noted, however, that although in the study of Winniford et al [87] high doses of ISDN were used, these were not the maximal doses recommended in cases of severe angina pectoris.

Bethge and Bachmann [88] compared the efficacy of sublingual ISDN, 10 mg, to that of sublingual (20 mg) and intracoronary (0.1 mg) nifedipine in 37 patients with suspected vasospastic angina. ISDN counteracted the effects of ergonovine on hemodynamics and left ventricular function in these patients, and increased the luminal diameter of the epicardial coronary arteries from 3.5 ± 0.8 to 4.9 ± 0.8 mm. Sublingual nifedipine did not alter the luminal diameter of the coronary arteries within 10 minutes after the ergonovine test. Intracoronary administration of nifedipine increased the coronary luminal diameter to the same extent as achieved by sublingual ISDN. Intracoronary nifedipine reduced aortic blood pressure and increased heart rate, but it did not alter the left ventricular pump function. Hence, sublingual ISDN was found as potent as intracoronary nifedipine in releasing the ergonovine-induced coronary constriction and more potent in counteracting the effects of ergonovine on left ventricular pump function.

Recently Distante et al [89] reported a preliminary study in patients with vasospastic angina, showing that oral IS-5-MN, 20 or 40 mg, is an effective therapeutic agent for prevention of myocardial ischemia due to vasospasm induced by ergonovine or isometric stress test. This study is discussed further in the chapter on mononitrates.

In summary, nitrates are effective in relief and prevention of vasospastic angina pectoris, within the well-known limitations of all forms of conservative medical treatment in this unstable condition. A controlled comparative study of nitrates and calcium antagonists in this condition is required.

7. Long-Term Use of Nitrates in Angina Pectoris

Several investigators have suggested that long-term treatment with nitrates in patients with angina pectoris is associated with development of tolerance. Although this theory is in contrast to the clinical experience of most cardiologists, it is supported by some evidence. In this chapter the subject is reviewed, with the conclusion that tolerance to the antianginal effect does not develop or can be reduced if the drugs are appropriately used. The topic is subdivided into three main areas:

1. vascular tolerance
2. tolerance to the antianginal effect
3. cross tolerance

The possible development of tolerance to the hemodynamic effects of nitrates in congestive heart failure is discussed separately in the chapter on congestive heart failure.

7.1. Does Tolerance to the Vascular Effect of ISDN Develop During Chronic Therapy?

Even with short-term administration the vasodilator action of nitrates is subjected to powerful counterregulatory reflex effects. With long-term use these mechanisms may result in attenuation of the effect of these drugs and the development of tolerance. This mechanism resembles that which occurs with other vasodilators. More frequently, however, nitrate tolerance is considered to result from an alteration in the target organs, namely in the vascular smooth muscle cells.

Several studies have shown partial vascular tolerance, with a differential rate of development and susceptibility of the venous and arterial vasculature [90, 91, 92]. The predominant site of vascular tolerance is controversial. Several investigators have reported that tolerance to the arterial dilator effect of nitrates develops most rapidly [92, 93] whereas others claim that tolerance is preferentially observed in the venous vasculature [94, 91]. One group has even suggested the development of arterial tolerance only.

A recent study on this topic was reported in 1986 by Stewart et al [95]. They found that long-term administration of nitroglycerin results in tolerance to its venodilatory effects, whereas the arteriolar dilator effect was maintained. As the initial effect of nitroglycerin is on the venous circulation, the development of preferential tolerance results in a shift in the site of action. However, the investigators found that this shift is normally marked by compensatory mechanisms, mainly those mediated by the autonomic nervous system. This study supports the report of Manyari et al [96] who observed that long-term nitrate treatment considerably attenuated the effects of nitroglycerin on vascular capacitance.

In another study, reported in 1987, Schneider et al [97] investigated the effect of ISDN on relaxation under isometric conditions of circular vessel strips of human saphenous veins obtained during cardiac surgery. They compared veins from patients not treated with organic nitrates prior to surgery to those of patients pretreated for more than 2 weeks with ISDN, 40, 80 and 160 mg daily. Contraction of the vessel strips was induced by sodium chloride, and ISDN was then added to the solution. In the venous vessel strips of all four groups comparable concentration-dependent relaxation was measured, irrespective of ISDN pretreatment. In conclusion, chronic pretreatment of patients with ISDN, even at high doses, does not impair relaxation of isolated circular venous vessel strips.

Other investigators have suggested that tolerance may not be related to changes at a certain vascular segment but to plasma volume expansion. Bennett et al [98] reported that administration of transdermal nitroglycerin to healthy human subjects resulted in plasma volume expansion in the presence of maintained vasodilatory effect (on abdominal vessels) and attenuation of hemodynamic response. Further studies are required to evaluate this mechanism.

All these studies evaluated the development of tolerance in the peripheral vasculature. It should be remembered, however, that nitrates also have a dilator effect on the coronary arteries, which probably contributes to their antianginal effect. Stewart et al [95] have observed tolerance to the dilator effect of nitroglycerin on large coronary arteries in dogs.

It is difficult to apply the findings of these studies to the clinical use of nitrates. The studies are not relevant for patients with congestive heart failure in whom many compensatory and regulatory mechanisms are maximally activated even before treatment. The relevance of these studies for patients without congestive heart failure is also limited because of variability in response. At very low doses, for example, only a venodilatory effect is observed. However, the shift in the spectrum of nitroglycerin action from a predominant venodilator toward an arteriolar dilator effect during long-term treatment should be considered when trials to evaluate nitrate tolerance are planned.

Perhaps the most important trial demonstrating sustained hemodynamic and anti-ischemic efficacy of ISDN in patients with angina pectoris was reported in 1986 by Wortmann et al [99]. They examined and compared the effects of initial and chronic (4-week) therapy with ISDN, 120 mg, sustained-release, in nine patients with coronary artery disease and impaired left ventricular function. At intraindividually identical workloads there was a reduction of pulmonary capillary wedge pressure from 32.5 ± 9.5 to 19.7 ± 9.8 mmHg. This reduction was fully achieved also during chronic therapy. With the first dose of ISDN cardiac index at maximum workload increased from 6.0 ± 1.2 to 6.8 ± 1.3 L min^{-1} m^{-2}. This hemodynamic improvement was associated with sustained improvement in exercise capacity and in ST-segment depression.

Recently Distante et al [100] used a different approach to study the possibility of tolerance to long-term nitrate therapy. They evaluated the hemodynamic response to an acute infusion of ISDN, 1 mg/min, up to a maximal dose of 10 mg, in eight patients with vasospastic angina chronically treated with high doses of transdermal ISDN, 150–300 mg daily. Each patient was studied before initiation of treatment and at regular intervals during chronic ISDN treatment for up to 36 months. The response to acute infusion of ISDN, including lowering of blood pressure and acceleration of heart rate, did not significantly decrease in relation to the duration of chronic treatment with transdermal ISDN. These results suggest that vascular tolerance does not develop even after long periods of transdermal treatment with ISDN.

In summary, although there is evidence for partial attenuation of the direct vasodilating effect of nitrates in certain segments of the circulation during chronic administration, it is most probable that this attenuation is not complete and does not affect all segments of the circulation. Thus the bulk of evidence indicates that vascular tolerance is only partial and segmental. The results of experimental studies on direct vascular effects cannot, therefore, be used to indicate frequent development of clinically significant tolerance to the therapeutic effect of nitrates.

7.2. Does Tolerance to the Antianginal Effect of ISDN Develop During Chronic Therapy?

As every clinician knows from his experience, ISDN is effective over years of treatment and without development of tolerance in many patients. This is supported by most clinical studies performed specifically to evaluate this aspect. However, some studies, and mainly those of Parker and coworkers [93, 101], suggest the development of at least partial attenuation of the antianginal effect of oral ISDN during chronic treatment.

Numerous studies indicate that tolerance does not develop. This was found to be true for clinical and, usually, also for hemodynamic parameters in patients with angina pectoris. In the present book these studies are described in detail in the chapter on the various nitrates used. Here a review of the most important papers is given.

Several controlled studies using various nitrate formulations showed sustained improvement of exercise tolerance in patients with angina pectoris [102, 103, 104]. Danahy and Aronow [303] reported that the reduction in systolic arterial pressure after a dose of ISDN persisted for a similar duration in acute and chronic therapy. Moreover, the increase in exercise tolerance persisted for 3 h after the dose in acute therapy and for 5 h during prolonged therapy. Niederer et al [105] reported that the decrease in left ventricular end-diastolic pressure and pulmonary arterial pressure and the increase in ejection fraction during exercise produced by ISDN in patients with angina pectoris were sustained throughout 4 weeks of treatment. A later report by Wortmann et al [99] (from the same group) emphasized that the hemodynamic improvement was associated with improvement in exercise capacity (from 414 to 686 w/min acutely and to 772 w/min chronically) and a decrease in ST-segment depression (measured as the sum of ST depression in all 12 standard leads, from 0.63 to 0.11 mV acutely and 0.16 mV chronically).

Schneider et al [67] reported that the antianginal effect of a high dose of ISDN (effects on exercise-induced ST-segment depression and frequency of anginal episodes) was sustained throughout 4 weeks of treatment. In a more recent study these investigators reported sustained efficacy without clinical tolerance also with conventional doses of ISDN [106]. A similar recent study with IS-5-MN also revealed no evidence of tolerance [107]. Even from Parker's laboratory, which had previously published convincing evidence of tolerance, a study which revealed only minor attenuation of the initial effect of ISDN during chronic therapy has been recently reported [93].

Nyberg and Nilsson [109] reported that IS-5-MN, in a controlled-release formulation, given once daily has an antianginal effect at 3 h after administration. The drug significantly increased total work load until onset of chest pain and until depression of ST-segment by 1 mm. The effect was maintained after 1 week of treatment with no evidence of development of tolerance to the antianginal and anti-ischemic effects.

Another important question is related to possible development of tolerance to intravenous nitrates in patients with coronary artery disease. Unverdorben et al [109a] have recently studied 22 patients with coronary artery disease who were randomly assigned to receive either 4 mg/h of ISDN or placebo, intravenously, and were given an additional dose of ISDN 10 mg or placebo, after 23 h. The patients were studied hemodynamically at rest and during 4 exercise tests. Placebo had no hemodynamic effect. ISDN reduced the pulmonary artery diastolic pressure by 42–49% at rest and by 29–37% during exercise. In 7 of the 12 patients receiving ISDN the initial effect was sustained while in 5 it was partially attenuated. No patient developed complete tolerance. Systemic blood pressure was contiuously lowered by ISDN. These findings indicate that even at a moderately-high infusion rate the hemodynamic effect of ISDN is sustained in most patients with coronary artery disease.

These studies, together with the clinical experience of most cardiologists, indicate a sustained antianginal efficacy of ISDN, without tolerance. Several other studies, however, suggest the development of at least partial tolerance. It is most important to simultaneously study clinical, electrocardiographic and hemodynamic effects. Such a trial has been reported by Ohlmeier et al [110] who performed serial cardiac catheterizations during bicycle exercise in patients with angina pectoris treated with various nitrate formulations for 2 weeks. The initial effect of ISDN, IS-5-MN and oral nitroglycerin on the response of pulmonary arterial pressure to exercise and on ST-segment depression was attenuated after 2 weeks of treatment.

Other studies claiming to demonstrate tolerance are less convincing. For example, Rennhak et al [111] reported that the initial effect of a slow-release formulation of ISDN on ST-segment depression was attenuated after 4 weeks of treatment. The effect of IS-5-MN, however, was largely sustained throughout 4 weeks of treatment.

Dalal et al [108] reported only partial hemodynamic tolerance. In ten patients with stable exercise-induced angina pectoris systolic blood pressure at rest decreased by 25% 2 h after acute administration of ISDN but by only 11% 2 h after administration of the drug during chronic therapy. Systolic blood pressure was not further decreased by sublingual nitroglycerin during the acute phase but was decreased further during prolonged therapy.

Various therapeutic regimens have been reported to be associated with tolerance. This has been shown with ISDN 15–120 mg [113], 80 mg ISDN slow-release twice daily [112], and 50 mg IS-5-MN three times daily.

Parker and coworkers have consistently reported that the antianginal effect of nitrates was abolished during continued treatment of up to 2 weeks [114, 115]. The most recent studies from this group, however, used long-acting formulations of nitrates, namely transdermal nitroglycerin disks and ISDN ointment, which might have special problems. The possibility of tolerance to transdermally administered nitrates is discussed separately in the chapter on this topic. Studies on long-term use of new nitrate formulations will clarify this point. Some preliminary studies with IS-5-MN suggest a sustained efficacy, without tolerance, although other studies indicate that mononitrates resemble other nitrates in this respect.

It is important to emphasize that even in the presence of vascular tolerance to nitrates the antianginal effect is sustained. This was demonstrated in a recent study by Kaltenbach and Schneider [116]. They observed a decrease or actual cessation of some of the cardiovascular effects of nitrates, including fall in blood pressure and rise in heart rate on standing, in ten patients with coronary artery disease during a 4-week period of treatment with ISDN, 20–40 mg six times daily. Evaluation of arteries and central veins by ultrasound confirmed the disappearance of the effect of nitrates on these vessels. Nevertheless, a crossover, double-blind, randomized trial demonstrated that the anti-ischemic effects of nitrates were sustained in patients. It was concluded that long-term treatment with ISDN maintains its antianginal activity even in the presence of diminution or complete abolition of the hemodynamic effects.

The question of tolerance to nitrates in ischemic heart disease is put in perspective by the study of low dose intravenous nitroglycerin in acute myocardial infarction reported by Jugdutt et al [116a] (described in detail in chapter 9). In this study significant hemodynamic tolerance was found in only 24% of the patients and in only 12% of them it was true tolerance. The vast majority of patients benefited, therefore, from the effect of nitroglycerin on ventricular function topography and complications.

7.3. Methods to Prevent Tolerance in Angina Pectoris

Tolerance has been attributed mainly to constantly high levels of nitrates in the plasma. Therefore it is reasonable to try to prevent the development of tolerance by regimens of intermittent administration, resulting in fluctuations in plasma levels of nitrates. This method is the principal means to prevent tolerance. Another method is the use of low-dose regimens, but it is limited by poor efficacy of very low doses.

One of the popular interval regimens consists of 20 mg ISDN twice daily, morning and midday. Blasini et al [117] reported that with this regimen the effects of ISDN on ST-segment depression, anginal frequency and sublingual nitroglycerin consumption were similar during acute and long-term therapy. It is not clear, however, whether the efficacy of this regimen results from the long interval between doses or from the low dose used. These investigators have also demonstrated that once-daily administration of slow-release ISDN, 120 mg, is not associated with tolerance. This is probably due to the fact that this formulation does not allow a build-up of constantly high plasma levels for 24 h.

Silber et al [112] compared in a randomized study two groups of patients with chronic stable angina pectoris who received ISDN, 80 mg, in an oral sustained release form, for 2 weeks, either once or twice daily. The anti-ischemic effect was evaluated by changes in exercise-induced ST-segment depression and left ventricular ejection fraction measured by radionuclide ventriculography. None of the ten patients who received the drug once daily showed any signs of attenuation of the anti-ischemic effect. In contrast, 7 of the 12 patients who received the tablets twice daily showed significant attenuation of the effect. These findings indicate that at these doses, and when a sustained-release formulation is used, once-daily administration is preferable to avoid tolerance. Similar findings were observed with IS-5-MN. Nyberg and Nilsson [109] reported that a controlled-release formulation of IS-5-MN had a sustained antianginal effect when given once daily. The effect was not sustained when given twice daily, possibly due to development of tolerance.

In a recent review Boertz and Bonn [118] suggested that sustained-release formulations might be an optimal solution to the problem of tolerance. On the one hand, it is now known that long-lasting, unfluctuating plasma concentrations of more than 300 ng IS-5-MN/ml cause a loss of action. The efficacy may be maintained by a regimen in which an interval with considerably reduced nitrate levels is guaranteed. On the other hand, a dosage regimen that achieves steady-state plasma concentrations which fluctuate between 100 and 300 ng IS-5-MN/ml (as occurs with conventional tablets of ISDN, 20 mg three times daily) showed no clinically significant loss of efficacy even after 4 weeks of use.

Based on these findings it was suggested that a therapy of once-daily 120 mg ISDN sustained-release, achieving plasma levels of up to 600 ng IS-5-MN/ml over the active daytime period of the patient, with an interval during which the plasma levels decline to 100 ng/ml overnight, could be effective. Recent studies demonstrate no loss of efficacy after 4 weeks of application. With regard to better compliance this therapy may be superior to conventional tablets three times daily.

Tolerance to transdermal nitroglycerin may also be prevented by using a nitrate-free interval at periods during which the drug is less clearly indicated (during the night for patients with exercise-induced angina pectoris). This is discussed further in the chapter on transdermal nitroglycerin.

7.4. *Cross-Tolerance to the Antianginal Effect of Sublingual Nitroglycerin in Patients Treated with Long-Acting Nitrates*

The question of cross-tolerance to nitroglycerin is of special significance when long-term nitrate therapy is considered. If long-term treatment with organic nitrates or transdermal nitroglycerin is associated with tolerance only to these agents, then the preventive effect is lost, but sublingual nitroglycerin is still effective for relief of pain in acute anginal episodes. If, however, cross-tolerance to nitroglycerin does develop, long-acting nitrates not only attenuate their own effect but also abolish the most effective measure of acute antianginal therapy.

From the experience of most clinicians and investigators it appears that nitroglycerin effectively relieves pain of angina pectoris even in patients on transdermal nitroglycerin or long-acting organic nitrates. Moreover, patients on long-term nitrate therapy are still responsive to acute administration of ISDN oral spray. This experience indicates that cross-tolerance does not develop.

There are several studies indicating the development of partial vascular cross-tolerance to nitroglycerin, but this is different from cross-tolerance to the antianginal effect (as vascular tolerance to nitrates differs from tolerance to the antianginal effect). Partial vascular tolerance may be observed in the presence of preserved antianginal effect. Moreover, it should be remembered that in all studies demonstrating partial cross tolerance, this has been demonstrated for only one hemodynamic parameter, such as systolic blood pressure. Such findings cannot be applied for the whole spectrum of cardiovascular effects of nitrates. This subject is discussed further in the section on vascular tolerance to nitrates.

Thadani et al [101] reported that in patients with chronic stable angina pectoris the first dose of ISDN, 15 mg, reduced systolic blood pressure by 36 mmHg. During long-term treatment a similar dose reduced systolic blood pressure by only 7 mmHg. The first dose of 0.6 mg nitroglycerin, before initiation of ISDN, reduced systolic blood pressure by 40 mmHg, whereas a similar dose during prolonged treatment with ISDN reduced systolic blood pressure by only 10 mmHg. Dalal et al [108] reported that in patients with stable exercise-induced angina pectoris sublingual nitroglycerin did not further reduce systolic blood pressure after acute administration of ISDN. During chronic treatment with ISDN, however, nitroglycerin did produce a further decrease in systolic blood pressure.

The relevance of these findings to the antianginal efficacy of nitroglycerin is not greater than the relevance of the well-known fact that the antihypertensive effect of nitroglycerin is rapidly attenuated, to the use of nitroglycerin in coronary artery disease.

Even in the study of Dalal et al [108], which showed partial vascular tolerance, no cross-tolerance to the antianginal effect of nitroglycerin was demonstrated. Nitroglycerin did not prolong exercise duration until angina in patients studied 6.5 h after the first dose of ISDN but it did prolong this period in patients studied, at the same interval, during chronic treatment with ISDN. This finding shows some attenuation of the effect of ISDN, but no cross-tolerance to nitroglycerin.

Naito et al [118a] attempted to evaluate the possibility of cross tolerance to sublingual nitrate in more than one segment of the circulation. This approach is important because the mechanism of action of nitrates involves more than one circulatory bed. They studied two groups of 12 patients each, with coronary artery disease. One group was on sustained ISDN therapy and the other did not receive long-acting nitrates. The definite evaluation included sublingual administration of IDSN, before and after which aortic systolic pressure, left ventricular end-diastolic pressure and coronary artery diameter were evaluated by cardiac catheterization. Sublingual ISDN produced less reduction of aortic systolic pressure and left ventricular end-diastolic pressure and less dilation of the coronary arteries in patients receiving sustained therapy with ISDN than in those without such therapy. However, a considerable effect was observed also in the patients receiving sustained therapy. These findings indicate that ISDN may be given sublingually for relief of angina even on patients who are on long-term treatment with this agent and, probably also with other nitrates.

7.5. Hemodynamic Effects of Bolus Nitroglycerin Compared with Those of Continuous Infusion

When administered intravenously, nitroglycerin is usually given in the form of continuous infusion. Most hemodynamic studies have been performed with this mode of administration. Recently it was shown that the hemodynamic effects of bolus injection of nitroglycerin in patients with coronary artery disease might be different from those of a continuous infusion. This difference is relevant to the understanding of the mechanism of the antianginal effect of nitroglycerin. Margolis et al [119] reported that observations on more than 1500 patients suggest that in the supine, properly hydrated patients intravenous nitroglycerin produces an initial rapid decrease in systolic, diastolic and mean blood pressure. During the subsequent 30–60 seconds blood pressure gradually returns to baseline. Left ventricular filling pressure initially falls, concomitantly with the systemic arterial pressure, but it continues to fall when the systemic arterial pressure returns to baseline levels. These findings suggest that unlike continuous infusion of nitroglycerin, which acts mainly on the capacitance vessels, bolus injection of nitroglycerin may act primarily on resistance vessels. Because of this different profile bolus nitroglycerin may have therapeutic uses different from those of continuous nitroglycerin infusion.

7.6. Mechanism of Vascular Tolerance to Nitrates

Tolerance to nitrates has been found to be associated with decrease in their primary effect – the enhancement of guanylate cyclase activity. Chronic exposure to nitroglycerin is directly related to down-regulation of the cGMP system. In animals which were made tolerant to nitroglycerin the activation of guanylate cyclase by the drug was reduced by about 75%, and the cGMP-phosphodiesterase activity was markedly increased [120]. In view of the cGMP effect Kukovetz et al [121] have recently tried to find out whether tolerance is a peripheral phenomenon occurring in the vascular smooth muscle, and whether it is related to an impairment of guanylate cyclase activation.

In isolated circular strips of coronary arteries 90-min pretreatment with nitroglycerin lowered the relaxation and the cGMP response to nitroglycerin, indicating induction of tolerance. Tolerance, although to a lesser extent, was also obtained with other organic nitrates. Minimal tolerance was found with nitroprusside, but no tolerance with SIN-1, the active metabolite of molsidomine. These two agents stimulated soluble guanylate cyclase in vitro even in the absence of cysteine, whereas organic nitrates required the presence of cysteine.

Preincubation with nitroglycerin, but not with other organic nitrates, almost completely inactivated guanylate cyclase. In conclusion, tolerance is related to impairment of guanylate cyclase function in the smooth muscle cells. The different degrees of tolerance development by various nitrovasodilators may be related to different mechanisms of activation and inactivation of guanylate cyclase, as well as differences in cysteine requirement.

The reduced effects of nitrates may result from decreased metabolism of these agents in the vascular walls. This metabolic process occurs prior to vasodilation. It was suggested that nitrosothiol, one of the products of this metabolism, may not be associated with tolerance.

Münzel et al [121a] studied the effect of the N-acetylcysteine (molecule shown to reverse the development of tolerance) on tolerant epicardial coronary arteries in chronically instrumented dogs. N-acetylcysteine did not restore the nitroglycerin responsiveness of these arteries. However, a small, tolerance-independent augmentation of the coronary dilatation produced by nitroglycerin was observed. It was found in dogs in whom tolerance was not induced. It may be attributed to extracellular formation of a guanylate cyclase stimulant from nitroglycerin, induced by N-acetylcysteine.

These investigators analyzed the literature according to the new hypothesis. They stated that the reported cases of enhancement of nitroglycerin effect by N-acetylcysteine did not demonstrate any specificity for tolerance. This could result from the fact that the transiently formed extracellular guanylate cyclase stimulant has an effect too weak to be detected in markedly tolerant blood vessels.

7.7. Effect of Nitrates on Angiographic Course of Coronary Artery Disease

There is recently interest in an attempt to delay the progression of coronary artery narrowing by the same drugs that are used to treat and prevent chest pain in patients with chronic stable angina pectoris. The theoretical background to this attempt is somewhat questionable. Beta-blockers may inhibit atherogenesis but, on the other hand, they have potential negative effects on plasma lipids. Calcium antagonists have an anti-atherosclerotic effect in animals. Nitrates are known to improve the patterns of coronary flow but have not direct anti-atherosclerotic effect. This background stimulated studies of the effect of these agents on coronary atheroslerosis in patients.

A group of 113 patients with chronic stable angina pectoris and proven coronary artery disease was treated with nifedipine, propranolol or ISDN for 2 years [121c]. They were assigned to the treatment groups after a comparative evaluation of the 3 drugs, by exercise test, during 3-days periods of treatment with each drug before initiation of the trial. The doses used were: nifedipine – 20 mg 4 times daily, propranolol – 80 mg 4 times daily, ISDN – 40 mg twice daily. Coronary arteriography was performed before and after 2 years of treatment.

The number of stenoses with evidence of progression was smaller with ISDN (24) than after propanolol (39). The result with nifedipine was even better. These findings indicate that nifedipine and, to some extent, also ISDN, are superior to propranolol in preventing the progression of established coronary artery disease.

The main limitation of this study is the dosage regimen. Nifedipine and propranolol were given at high doses. ISDN was used, however, according to a regimen that may be associated with tolerance. It might have been better to use a slow release ISDN which is not associated with tolerance. Moreover, no attempt has been mad to evaluate the possibility of tolerance by repeated exercise tests.

Despite the limitations of this study the approach offer important potential advantages. Further controlled clinical trials of the effect of nitrates and other anti-anginal drugs on coronary atherosclerosis are indicated.

8. Silent Ischemia

The concept of silent ischemia has added a new and important indication to nitrate therapy. It is beyond the scope of this book to discuss silent ischemia.

Two points should, however, be emphasized:

1. Silent ischemia is the result of mechanisms similar to those resulting in painful ischemia, i.e., stable and unstable angina pectoris. The extent, time course and interaction of these mechanisms may differ between angina pectoris and silent ischemia, but basically they resemble each other.
2. Silent ischemia impairs prognosis of patients with coronary artery disease and therefore should be treated. Naturally, nitrates were among the first agents evaluated for treatment of silent ischemia. Their investigation was intensified after it was realized that episodes of silent ischemia are often not preceded by acceleration of heart rate, and that beta-blockers may therefore not be effective.

Nitrates, in monotherapy or in combination with other anti-ischemic agents, have been found effective in reducing the number of episodes of silent ischemia.

My colleagues and I have presented at the symposium on „Advances in the Clinical Practice of Nitrate Therapy" (Montreux 1987) the results of a preliminary trial in which transdermal nitroglycerin patches were given to patients with angina pectoris who had silent ischemia – despite treatment with beta-blockers in all and calcium antagonists in some. The nitroglycerin patches reduced the number of silent ischemic episodes by over 80%. This is discussed further in the chapter on transdermal nitroglycerin.

At the same symposium (to be published in The American Journal of Cardiology) von Arnim and Erath reported a controlled double-blind trial comparing the effects of IS-5-MN and nifedipine in 12 patients with documented transient silent ischemia. Seventy percent of ischemia episodes were unaccompanied by pain.

The patients received in four weekly periods IS-5-MN, 20 mg three times daily or 50 mg slow-release once daily, or nifedipine, 20 mg slow-release tablets three times daily. At the end of each week Holter monitoring was performed, showing reductions in episodes by 68% and 65% with IS-5-MN and 65% and 68% with nifedipine. Painful and painless episodes were reduced in similar extent. Individual responses showed great variability, and in all treatment periods not more than half the patients became completely free of ischemia episodes. One of 12 patients did not respond to either method of treatment. It was concluded that both IS-5-MN and nifedipine reduce ischemic episodes by about 65%. A much longer study, evaluating the effect of transdermal nitroglycerin on silent ischemia by eight periods of Holter monitoring in each patient, is now being performed by my group. The most important question to be answered is whether the beneficial effect of transdermal nitroglycerin in silent ischemia is maintained throughout chronic therapy.

Feng et al [121b] studied the efficacy of IS-5-MN, 20 mg 3 times daily, on silent myocardial ischemia after myocardial infarction in 28 Chinese patients, evaluated by 48 h ambulatory electrocardiographic monitoring in a randomized crossover single blind placebo-controlled study. The drug significantly reduced number of painful and silent ischemic episodes compared with placebo. The number of total ischemic episodes was reduced by 88%, their duration – by 94%, the time-ischemia integral by 95% and the total ST maximal depression – by 86%. The drug did not alter heart rate or blood pressure.

In conclusion, nitrates are effective monotherapy for silent ischemia in patients with various presentation of coronary artery disease.

9. Acute Myocardial Infarction

In previous chapters it has been shown that nitrates are used to treat various preinfarction ischemic conditions as well as postinfarction complications, such as acute heart failure, mechanical complication and postinfarction angina. Nitrates may also be used for acute myocardial infarction per se. Moreover several early intervention studies revealed beneficial effects of other anti-ischemic agents such as beta-adrenoreceptor blockers in acute myocardial infarction. Therefore it is interesting to review the facts and hypotheses on the use of nitrates in these conditions.

In "complete" myocardial infarction without complications nitrates theoretically have no potential role. This view, however, does not take into account the dynamic nature of acute myocardial infarction. The terminal event leading to infarction is impairment of the myocardial oxygen supply/demand ratio. This ratio may be altered in either direction even after myocardial necrosis has begun. Increased flow to the ischemic area, either from collateral vessels or by dilation of the critical stenoses which caused infarction, or decrease in myocardial oxygen demand may reverse the ischemic process and preserve viability of the myocardium.

In discussing the use of anti-ischemic agents in acute myocardial infarction it should be remembered that interventional cardiology has changed the classical course of acute myocardial infarction. Thrombolysis, percutaneous transluminal coronary angioplasty and early surgical revascularization have made most pharmacologic, early-intervention studies irrelevant, as in few patients does myocardial infarction follow its natural pattern. Therefore, new secondary prevention studies are required, not after acute "complete" myocardial infarction but after successful interventions preventing the development of such an event.

9.1. Effects of Nitrates on Cardiovascular Function in Acute Myocardial Infarction and Preinfarction

Animal and human studies have revealed that in acute myocardial infarction nitrates reduce left ventricular filling pressure and, to a lesser extent, also systemic vascular resistance, and that they increase cardiac output, coronary blood flow and contractility of ischemic segments and, in some cases, decrease infarct size. The effect shows marked interpatient variability, depending both on the physiologic and hemodynamic condition and on coronary anatomy.

The decrease in left ventricular filling pressure, resulting from preload reduction and anti-ischemic effect, is probably the most prominent effect of nitrates in acute myocardial infarction [122–124]. This effect is greater in patients with high pretreatment levels of left ventricular filling pressure, but it has also been observed in patients with normal pretreatment levels. A significant decrease in systemic vascular resistance has usually been observed only with moderate to high doses of nitrates. It is important not to reduce systemic arterial pressure to levels too low to impair coronary perfusion. Some authorities give nitrates when systolic pressure is as low as 100 mmHg.

In patients with acute myocardial infarction and normal left ventricular filling pressure nitrates can reduce cardiac output, whereas in those with elevated filling pressure nitrates increase cardiac output. The increase in cardiac output, which is most prominent in patients with heart failure, results from afterload reduction, improvement of myocardial contractility due to the anti-ischemic effect, and acceleration of heart rate.

Nitrates cannot, of course, alter contractility of necrotic akinetic myocardial segments at the center of the infarct zone. They can, nevertheless, enhance contractility of hypokinetic segments in the periphery of the ischemic zone [125–128]. Nitrates have also been shown to reverse the depressant effect of propranolol on both ischemic and nonischemic myocardial segments after acute coronary occlusion. Shimoura et al [122] reported that the extent of enhancement of myocardial contractility was inversely related to the extent of ischemic damage. Segments with over 60% necrosis showed no potentiation of contractility with nitroglycerin whereas segments with less than 40% necrosis showed significant potentiation.

Rezakovic and Rutishauser reported at the World Congress of Cardiology (Moscow 1982) on different hemodynamic effects of nitroglycerin and ISDN in acute myocardial infarction. An invasive hemodynamic study in 30 patients revealed that nitroglycerin has mainly a peripheral venodilatory effect, and that it also reduces pulmonary vascular resistance. ISDN produces these effects and, in addition also reduces the systemic vascular resistance.

Bussmann and coworkers (Frankfurt/M.) published over 10 years ago a series of reports on the hemodynamic effects of various nitrate formulations in acute myocardial infarction. In ten patients with acute myocardial infarction and five with old infarction sublingual nitroglycerin, 0.8–1.6 mg, significantly lowered the left ventricular filling pressure and pulmonary arterial pressures within 3–5 min after administration. In six of the patients who had acute pulmonary edema the dyspnea was relieved within a few minutes after administration of nitroglycerin sublingually.

Another group of 24 patients with acute myocardial infarction received nitroglycerin intravenously, 3 mg in the 1st hour and 6 mg in the 2nd. They were divided into two groups, with left ventricular filling pressure, respectively, above or below 20 mmHg. In both groups the drug reduced left ventricular filling pressure, from 28 to 16 mmHg and from 15 to 9 mmHg, respectively. Cardiac output decreased from 4.4 to 3.9 L/min in the group with low filling pressure and increased from 3.5 to 4 L/min in the group with high filling pressure. The size of the ischemic zone, determined by ST-segment deviation, was reduced by the lower infusion rate.

A similar study was performed in 21 patients with acute myocardial infarction, with or without left ventricular failure, treated with 10 and 20 mg ISDN orally. In those patients with left ventricular filling pressure of less than 20 mmHg this pressure was reduced from 13.6 ± 4 to 7.1 ± 2.6 mmHg. In those with pretreatment filling pressure over 20 mmHg it was reduced from 26.9 ± 4.6 to 19.0 ± 3.6 mmHg. Cardiac output decreased in the first group and increased in the second. Systemic arterial pressure decreased slightly but not significantly in both groups. The optimal hemodynamic profile is a decrease in left ventricular filling pressure with maintenance of the coronary perfusion pressure. This may be achieved by a combination of nitrates and intra-aortic balloon counter-pulsation. This method was first described by Rudolph and his coworkers at the symposium on "The First 24 h of Acute Myocardial Infarction" in 1977.

Rabinowitz and coworkers have shown that nitrates can improve cardiovascular function also in patients with myocardial infarction and hyperdynamic circulation. In these patients increased production of catecholamines results in tachycardia and systemic hypertension. Fifteen such patients with heart rate over 90 beats/min, diastolic blood pressure over 90 mmHg and pulmonary capillary wedge pressure over 15 mmHg were studied before and during intravenous infusion of ISDN, 2–8 mg/h. The level of plasma catecholamines was reduced by about 75%, heart rate decreased from 118 to 88 beats/min, and diastolic blood pressure decreased from 100 to 76 mmHg.

Evaluation of intravenous ISDN and nitroglycerin in the treatment of acute myocardial infarction with elevated pulmonary capillary wedge pressure ($\geqq 18$ mmHg) was performed by Cintron [129] in a group of 111 patients in a single-blind multicenter study. The patients were randomly assigned to receive either ISDN or nitroglycerin at doses titrated until a 25% decrease in pulmonary capillary wedge pressure was reached. This was achieved in 82% of the patients with ISDN and in 92% with nitroglycerin. In the presence of a 35% reduction in pulmonary capillary pressure with ISDN cardiac output increased by 9% and systolic blood pressure was reduced by 6%. With nitroglycerin capillary pressure was reduced by 36%, cardiac output increased by 17%, and systolic blood pressure reduced by 4%. Heart rate was not altered with either drug. Patients who were stabilized on ISDN required no significant dose changes during a maintenance phase of 24–48 h. In contrast, patients receiving nitroglycerin required progressively higher doses. The most common adverse effect was hypotension – in 14% of the patients on nitroglycerin and 8% on ISDN.

9.2. Effect on Size and Complications of Acute Myocardial Infarction

The anti-ischemic effects of nitrates can result in a decrease in infarct size. This has been demonstrated in experimental and clinical studies, using pathologic, enzymatic and electrocardiographic evaluation of infarct size [130–137]. In many of these studies it was shown that maintenance of adequate coronary perfusion pressure is essential for the beneficial effect. Systolic arterial pressure should not be lowered beyond 90 mmHg. The decrease in infarct size was associated with clinical improvement.

Chiche et al [138, 139] and Derrida et al [140] used an electrocardiographic analysis of the QRS complex to evaluate infarct size. They found a significant decrease in infarct size and in the frequency of early sudden death in patients treated with nitroglycerin. These investigators attributed the beneficial effect to early treatment of left ventricular failure.

The leaders in the use of nitrates in acute myocardial infarction are Bussmann and coworkers from Frankfurt/M. [141]. They measured infarct size by the enzyme CK and the isoenzyme CK-MB in 60 patients, half of them treated with nitroglycerin intravenously. Infarct size was reduced by 30% in the treated group compared with the control group. This effect was observed in both early and late intervention trials (up to 8 h from onset of symptoms in the early intervention group and a mean interval of 12.8 h from symptoms in the late intervention group). The patients treated with nitroglycerin had less left ventricular failure, chest pain, morphine requirement and ventricular and supraventricular arrhythmias than the control patients. Moreover, patients in the control group had more reinfarctions than the patients treated with nitroglycerin.

Kim and Williams [142] reported that multiple sublingual doses of nitroglycerin abolished chest pain and reduced electrocardiographic signs of myocardial damage in patients with acute myocardial infarction.

Partially different results were found by other investigators. Jaffe and coworkers [143] reported that intravenously administered nitroglycerin, given up to 10 h from onset of symptoms, reduced enzymatically evaluated infarct size in patients with inferior but not anterior acute myocardial infarction. Flaherty et al [144] reported a placebo-controlled study of intravenous nitroglycerin infused for 48 h, in 104 patients with acute myocardial infarction. In the patients in whom treatment was initiated within 10 h from appearance of symptoms treatment was associated with a reduced incidence of congestive heart failure, infarct extension and cardiac death compared with the placebo group. Left ventricular ejection fraction was increased by more than 10% in 35% of the patients treated early with nitroglycerin compared with 6% of those treated late. No effect on infarct size was observed. In the patients treated after more than 10 h from onset of symptoms no effect on infarct size or on complications was observed.

It is not clear whether the anti-ischemic or the hemodynamic effects of nitroglycerin predominate in decreasing infarct size and complications. I believe that both effects play a role. Animal studies by Sheehan and coworkers [145] revealed, however, that nitroglycerin salvaged ischemic myocardium only in the presence of congestive heart failure. This suggests that the hemodynamic effects predominate. In most centers nitrates are not used at present for the indication of myocardial salvage. They are, however, frequently used to treat congestive heart failure in patients with acute myocardial infarction and probably contribute to limitation or even reduction of infarct size in this important subgroup.

One of the leading groups in research on nitrates in acute myocardial infarction is that of Maseri and coworkers at the Hammersmith Hospital in London. These investigators found that myocardial infarctions show, in the first 24–48 h, a tendency to extension of infarct size which might be considered "early reinfarction." This is evident in most patients as an appearance of a second peak of activity of the MB fraction of the enzyme creatine kinase. Since the introduction of systematic therapy with intravenous ISDN from the early stages of acute myocardial infarction this second peak has largely disappeared (provided that adequate measures, such as raising the legs and treatment with sympathomimetic amines, are taken to maintain adequate systemic blood pressure).

Grosser et al [146] studied, in a controlled, randomized trial the effect of treatment with ISDN, 60 mg orally, in 509 patients with acute myocardial infarction. A subgroup of 132 patients was treated within 8 h after onset of symptoms. Mortality in the whole ISDN group was not significantly different from that in the control group. However, mortality in the subgroup treated within 8 h from symptoms was significantly lower than that of the control group. In these patients, and particularly in those with elevated left ventricular filling pressure, the drug reduced the filling pressure and increased cardiac output. A significant decrease in arrhythmias and angina pectoris was found in the treated group. Hemodynamic improvement was observed when the patients treated within 8 h were evaluated before discharge from hospital.

Sclarovsky et al [147] studied the effect of high doses of ISDN, given intravenously within a short time, on the very early electrocardiographic stages of acute anterior wall myocardial infarction. Doses of 90 ± 25 mg, over 1 h, were used in patients with acute anterior wall myocardial infarction, showing ST elevation with tall-peaked T wave, without or with only minimal R wave. The dose of ISDN was titrated not to lower systolic blood pressure less than 100 mmHg and not to accelerate heart rate to over 100 beats/min.

In patients with no significant alteration of QRS configuration in leads V_2–V_3 there was no effect. In those with disappearance of S wave and increase of R wave in V_2–V_3 the ST-segment elevation was reduced from 14 to 5 mm after 1 h, the height of the tall-peaked T wave was reduced by 50%, left ventricular ejection fraction increased by 6%, and the left anterior descending coronary artery remained open.

It was concluded that in patients with this specific electrocardiographic pattern ISDN, at high doses, is an alternative treatment to thrombolytic drugs when these are contraindicated, especially in the elderly.

It should be mentioned, however, that the electrocardiographic classification of ischemia proposed by Sclarovsky and his coworkers is still under investigation and is not widely accepted. Further research is required to validate their conclusions.

Another large-scale clinical trial was reported in 1988 by Jugdutt and Warnica [147a]. Three-hundred and ten patients with acute myocardial infarction were randomly allocated to nitroglycerin and control groups. Nitroglycerin was given intravenously, at doses titrated to lower mean blood pressure by 10% in normotensive and 30% in hypertensive patients, but not below 80 mmHg. The infusion was maintained for 39 h. Infarct size, determined by creatine kinase measurements was less in the nitroglycerin than in the control group. This was found in both anterior and inferior infarctions. The effect was greater in those who received nitroglycerin early than in those who received it late.

Other indexes of infarct size and ventricular function were also improved by nitroglycerin. Ten days after the infarction left ventricular asynergy was 40% less in the nitroglycerin group compared with the control group, left ventricular ejection fraction was 22% more and Killip class score was 41% less. Expansion index increased by 31% and thinning ratio decreased by 17% in the control group at 10 days, but remained unaltered in the nitroglycerin group. Major complications, including infarct expansion syndrome, left ventricular thrombus, cardiogenic shock and infarct extension were less frequent in the nitroglycerin group. Mortality was reduced in the nitroglycerin group in hospital and at 3 and 12 months. This was attributed only to the subgroup with anterior infarction.

It may be concluded that in patients with acute myocardial infarction nitroglycerin therapy limits the infarct size, expansion and complications, independent of infarct location. The benefit is greater in patients who receive the drug early, and in whom mean blood pressure is maintained at or above 80 mmHg [116a].

9.3. Does Tolerance Develop to the Effect of Nitrates in Acute Myocardial Infarction?

As in the other indications for nitrate therapy, appropriate and "smart" use can result in a sustained beneficial effect of nitrates in the great majority of patients with acute myocardial infarction.

Jugdutt et al [116a] evaluated the question of vascular tolerance in 154 patients with acute myocardial infarction who were treated with continuous infusion of low-dose nitroglycerin, in a randomized placebo-controlled study. The infusion rate of nitroglycerin was titrated to lower mean blood pressure by 10% in the normotensive patients and 30% in the hypertensive patients, but not below 80 mmHg. Tolerance was defined as the need to increase the infusion rate to maintain the initial effect on blood pressure.

Throughout the nitroglycerin infusion left ventricular asynergy, dimension and volume were less, and left ventricular ejection fraction was greater compared with baseline or placebo. Thus tolerance did not make a major impact on grouped data. Reversal of blood pressure and cardiac volumes after discontinuation of nitroglycerin also suggests lack of significant tolerance in this group.

Individual analysis revealed that the initial effect was sustained in 76% of the patients, whereas in 24% there were some evidence of tolerance. However, in only half ot these patients, 12% of the group, this was true tolerance. In the remaining increases in nitroglycerin dose were required because of recurrent ischemic episodes that further impaired cardiac function.

This figure of 12% true tolerance to nitroglycerin is concordant with our experience with intravenous isosorbide dinitrate and it puts the question of tolerance in its appropriate perspective.

However, patients who developed tolerance had similar beneficial effects on left ventricular function and topography as those who did not develop tolerance. This might have resulted from the fact that tolerance became significant only after 10 h of treatment [116a].

Inappropriate use may result in a higher frequency of tolerance. The potential mistakes here are similar to those made in other indications for nitrate therapy, the most important being use of inadequately high doses.

9.4. Nitrates Combined with Thrombolysis in Acute Myocardial Infarction

The combination of nitrates with thrombolytic therapy may improve management of acute myocardial infarction. Hackett et al [148], summarizing the experience at Hammersmith Hospital, have shown that spontaneous intermittent coronary occlusion is frequently found both before and during the early phase of acute myocardial infarction. Coronary reocclusion during continuous intracoronary thrombolytic therapy was found to be responsive to ISDN. This finding suggests that vasoconstriction interacts with a thrombus to contribute to coronary occlusion in acute myocardial infarction. ISDN, in addition to interfering with this interaction, also improves delivery of thrombolytic agents to the thrombotic site. Therefore the addition of ISDN, and probably also other nitrates, to thrombolytic therapy has a definite role in improving coronary flow and reopening coronary reocclusion in patients with acute myocardial infarction.

Flaherty [149] recommended the use of intravenous nitroglycerin in conjunction with thrombolytic therapy or angioplasty in patients seen within the first 4 h of onset of symptoms and the use of nitroglycerin alone in patients seen 4–12 h after onset of symptoms or in patients seen earlier, in whom thrombolytic therapy is contraindicated. In my opinion this recommendation applies also to ISDN and IS-5-MN.

9.5. Secondary Prevention After Acute Myocardial Infarction

Several anti-ischemic agents have been shown to reduce mortality and morbidity in patients after acute myocardial infarction. This was shown in late and/or early intervention studies with beta-blockers (propranolol, metoprolol, timolol) and calcium antagonists (diltiazem). An attempt to use nitrates for secondary prevention appears promising. The anti-ischemic effect of nitrates may be especially beneficial in early-intervention studies. A large secondary prevention trial of nitrates is now planned in South America.

It should be remembered, however, that the only beta-blockers which fail to decrease mortality after acute myocardial infarction are those with sympathomimetic activity (alprenolol, oxprenolol). This has been attributed to inadequate decrease of heart rate by these agents. The main determinant of efficacy of beta-blockers in secondary prevention appears to be a reduction in heart rate. This suggests that acceleration of heart rate by nitrates might prevent expression of their anti-ischemic effect. However, the mechanism of action of nitrates differs from that of beta-blockers, and it is difficult to apply findings from studies of beta-blockers to studies of nitrates.

9.6. Effect on Postinfarction Clinical Course

The present state of knowledge allows only speculations on the effect of nitrates, given to patients with acute myocardial infarction, on the late clinical course.

The decrease in infarct size may result in reduction of late mortality. In contrast, salvage of ischemic myocardium may result in late arrhythmias originating in this zone. Bussmann et al [150] reported that five patients with myocardial infarction not treated with nitroglycerin died, compared with no deaths in a group of patients treated with nitroglycerin. However, the total number of patients in this study (60 patients) was too small to permit drawing statistically significant conclusions. Patients with myocardial infarction treated with nitroglycerin had a higher incidence and a greater severity of postinfarction angina pectoris after discharge from hospital. This probably represents salvage of border-zone ischemic myocardium, which is, of course, beneficial. These patients should undergo early coronary catheterization, percutaneous transluminal angioplasty or coronary bypass grafting.

The planned secondary prevention study will contribute to better understanding of the late postinfarction clinical course.

9.7. Nitrates in Acute Myocardial Infarction – Recommendations for Clinical Practice

At present there are no strict recommendations for the use of nitrates in acute myocardial infarction. However, the bulk of evidence supports the use of rapidly acting nitrates when acute myocardial infarction is first suspected. In the past clinicians used this measure to a limited extent, as they were concerned over the potential decrease in coronary perfusion pressure. At present such concern is much less relevant due to the availability of various tools to maintain the perfusion pressure. At present many clinics with facilities for hemodynamic monitoring, as well as clinics without such facilities, give nitrates from the very early stages of acute myocardial infarction. For ISDN and nitroglycerin clinical monitoring is sufficient in patients who do not have an unfavorable initial response. It is preferable, however, to perform hemodynamic studies, particularly at the initiation of treatment or when intravenous treatment is replaced by sublingual or oral therapy.

Another question is whether patients who are asymptomatic immediately after the onset of acute myocardial infarction should receive nitrates. It is known that reduction of infarct size by nitrates may lead to late development of angina pectoris and arrhythmias, originating in the salvaged myocardial zone. This requires, of course, long-term treatment with nitrates (or interventional measures). It should be remembered also that preliminary studies have revealed a decrease in mortality in patients with acute myocardial infarction treated with nitrates from the early stages. For the indication of secondary prevention the combination of beta-blockers and nitrates might be especially beneficial.

In summary, myocardial infarction, which in the past was a contraindication to nitrate therapy is now a clear indication for this therapy. This is mainly valid when the infarction is complicated by acute or chronic heart failure or post-infarction angina.

9.8. *The Relevance of the Antiplatelet Effect of Nitrates in Acute Myocardial Infarction and Unstable Coronary Syndromes*

The clinical relevance of the interaction between platelets, arterial damage, the coronary tone and the effect of nitrates is related to the understanding of the role of platelet activity in the pathophysiology of unstable coronary syndromes such as acute myocardial infarction, sudden death and unstable angina. The mechanisms of these syndromes include atherosclerotic plaque rupture, platelet aggregation and thrombotic and vasospastic coronary occlusion. Nitrate therapy has been beneficial in these conditions, as discussed in other chapters of this book.

The effect of nitrates on platelet function is described in chapter 4. Recently a specific experimental study of the interaction of platelets, vasoconstriction and a nitrate during arterial wall injury has been reported by Lam et al. [150a]. They found that in the presence of deep arterial wall injury nitroglycerin, administered intravenously at a dose which lowered mean arterial pressure by 9 ± 2% significantly decreased platelet deposition and the vasoconstrictive response. In the presence of mild arterial wall injury nitroglycerin reduced the vasoconstrictive response without affecting the already low level of platelet deposition. The reduction of platelet deposition after deep arterial injury may explain, at least in part, the beneficial clinical effect of nitrates in unstable coronary syndromes.

10. Congestive Heart Failure

Since it was realized that vasodilation might be beneficial in congestive heart failure [151], nitrates have become the cornerstone of this therapeutic modality. Even at low doses nitrates produce venodilation, resulting in decreases in left and right ventricular filling pressures. At higher doses nitrates cause, in a majority of patients, a decrease in afterload, resulting in an increase in cardiac output. Nitrates are effective both in acute and chronic congestive heart failure, and they may be used, usually as combination therapy, in the extreme presentations of this condition – pulmonary edema and cardiogenic shock. The hemodynamic response to nitrates is often associated with symptomatic improvement, particularly in symptoms related to congestion but also in those resulting from inadequately low cardiac output.

From the studies of Cohn and his coworkers [152–155] performed over 10 years ago it became evident that all nitrates produce a similar hemodynamic response, including decreases in pulmonary capillary and pulmonary arterial pressures, an increase in cardiac output, a small decrease in systemic arterial pressure and a slight acceleration of heart rate. This is further discussed in the section on hemodynamic effects. As might be expected, the various agents studied differed pharmacokinetically, resulting in different time courses in their effects. These differences stimulated an effort to optimalize treatment either through the development of new agents or through new formulations of administration.

At present nitrates are usually given to patients with chronic congestive heart failure who do not respond adequately to digitalis and diuretics. In most patients nitrates are used in combined therapy with digitalis and diuretics. In some patients in whom digitalis glycosides are clearly ineffective nitrates are used only with diuretics. Occasionally arterial dilators, such as hydralazine, are used in combination with nitrates.

One of the questions which at present remains unanswered is whether nitrates reduce mortality in patients with congestive heart failure. The results of the Veterans Administration Cooperative Study (V-HeFt) which has just been completed revealed a reduction in mortality.

Hemodynamic studies in congestive heart failure have been performed with nitrates given intravenously, orally, sublingually, transcutaneously and as an oral spray. These studies have produced comparable results.

Nitrates have been studied as monotherapy or in combination with digitalis, diuretics and other vasodilators. Comparative studies of nitrates with other vasodilators such as hydralazine and captopril have also been reported.

10.1. Hemodynamic and Clinical Effects

Acute administration of nitrates produces hemodynamic improvement in patients with acute and chronic congestive heart failure. This was observed with all nitrates and all routes of administration. Most studies, however, have been performed with nitroglycerin or ISDN, given intravenously.

The most consistent effect of nitrates in patients with congestive heart failure is reduction in left ventricular filling pressure. This reduction is usually by 10%–50%. Pulmonary arterial pressure and systemic systolic pressure are also reduced. Diastolic systemic arterial pressure is less consistently reduced. Mean right atrial pressure is also reduced. Systemic vascular resistance is reduced by nitrates, but usually at doses higher than those required to reduce left ventricular filling pressure (because veins are more sensitive than arteries to nitrates). In patients with very high levels of systemic vascular resistance this parameter may be reduced, however, even by low doses of nitrates. Heart rate may increase as a compensatory response to vasodilation, not change, or even decrease if stroke volume significantly increases or as a response to overall hemodynamic improvement.

Rabinowitz et al [156] and Baligadoo et al [157] reported that the increase in cardiac output is related to the pretreatment level of systemic vascular resistance. Intravenous ISDN increased cardiac output by 25% or more only in patients with systemic vascular resistance greater than 2000 dyn sec cm^{-5}. Patients with pretreatment systemic vascular resistance lower than 1500 dyn sec cm^{-5} showed no significant increase in cardiac output.

An increase in cardiac output can be observed only when left ventricular filling pressure is not excessively decreased. Williams et al [158] found a significant increase in cardiac output only in patients in whom left ventricular filling pressure remained above 12 mmHg after administration of nitrates.

Nitrates have a direct effect on the pulmonary vasculature. They can reduce pulmonary vascular resistance in patients with, as well as without, congestive heart failure.

These hemodynamic changes have been observed both in patients with chronic heart failure and in those with acute heart failure complicating acute myocardial infarction, both at rest and during exercise [159, 160]. The decrease in left ventricular filling pressure was greater at rest [161]. Ejection fraction increased at rest but was only minimally increased or not altered during exercise [162].

Similarly to other vasodilators, nitrates are not effective in all patients with congestive heart failure. About 15%–25% of patients with chronic congestive heart failure do not respond to acute administration of nitrates. In another 15%–20% of patients initial improvement is followed by late attenuation of the effect. (This response is discussed in detail in other chapters). It is more difficult to evaluate the number of nonresponders in acute pulmonary edema, as nitrates are not used alone in this condition. It is estimated not to be higher, and perhaps even lower, than the percentage of nonresponders in chronic congestive heart failure.

It is also important to evaluate the effects of nitrates on hemodynamic response to exercise. This is because the hemodynamic burden induced by exercise improves the capacity to correlate hemodynamic measurements to clinical status and prognosis in patients with chronic congestive heart failure. Franciosa and Cohn [163] evaluated the effect of acute administration of ISDN on the response to submaximal and maximal exercise in patients with congestive heart failure. The drug lowered left ventricular filling pressure and increased cardiac output at rest and submaximal exercise but not at maximal exercise. Maximal exercise capacity was not increased. The investigators suggested that as exercise produces vasodilation, the effect of vasodilators may be less during exercise than at rest. In a later study these investigators [164] revealed that maximal exercise capacity increased after 3 months of treatment with ISDN.

Nitrates produce only slight changes in the arterial blood flow to the limbs. Wilson et al [165] have demonstrated that, despite the overall hemodynamic improvement produced by ISDN during exercise in patients with chronic congestive heart failure, the drug did not increase skeletal muscle blood flow or oxygen availability. Renal vascular resistance was not altered. Renal blood flow may decrease as systemic arterial pressure is reduced. Hepatic vascular resistance and blood flow were usually not altered. Interestingly, in healthy volunteers nitroglycerin, but not ISDN, was shown to reduce the apparent liver blood flow. There is no explanation for this difference. It should be remembered, however, that in patients with liver diseases it is preferable to reduce portal pressure without reducing liver blood flow [166].

Effects of nitrates on duration of exercise in patients with chronic congestive heart failure are controversial. Hecht et al [167] reported that ISDN increased exercise duration by 28%, whereas Franciosa and Cohn found no change in duration of exercise.

The hemodynamic improvement produced by nitrates is usually associated with symptomatic improvement. This is most prominent in patients with acute pulmonary edema. It was shown that nitrate monotherapy (even other than by the intravenous route) can completely reverse acute pulmonary edema. Marmor et al [168] have shown this with a single dose of ISDN spray and Bussmann and Schupp [169] with multiple doses of sublingual nitroglycerin.

The study by Bussmann and Schupp [169], reported in 1978, was the first to show that hemodynamic and symptomatic improvement may be achieved in patients with acute pulmonary edema by administration of nitrates not by the intravenous route. In seven such patients hemodynamic studies were performed. Within 10 min of sublingual administration of nitroglycerin, 1.6 mg, left ventricular filling pressure decreased from 33 ± 10 to 24 ± 8 mmHg, and cardiac output increased from 3.3 ± 0.8 to 3.7 ± 0.8 L/min. In one patient with recent myocardial infarction left ventricular filling pressure decreased from 50 to 27 mmHg within 5 min. Dyspnea, orthopnea and pulmonary rales rapidly disappeared in these patients.

Another 15 patients with pulmonary edema were evaluated only clinically. They received one to six doses of 0.8–2.4 mg of nitroglycerin sublingually at intervals of 5–10 min. Five minutes after administration the first evidence of clinical improvement was observed in seven of the patients. After 15–20 min pulmonary rales disappeared or decreased in 11 and dyspnea decreased in 14 of the patients. The elevated systemic arterial pressure and heart rate decreased significantly. Only one patient with end-stage pulmonary edema remained nonresponsive to nitroglycerin.

Despite these impressive results it should be remembered that it is difficult to give several doses of sublingual nitroglycerin to distressed patients with pulmonary edema. They may be unable to hold the tablet or may swallow it. Marmor et al [168] solved this problem by using ISDN oral spray. This study is discussed in the chapter on nitrate oral spray.

It has been suggested that peripheral edema may reduce venous compliance in patients with congestive heart failure, thereby impairing the response of nitrates. Magrini and Niarchos [170] evaluated the effect of sublingual nitroglycerin, 1.6–2.4 mg, in 15 patients with acute left ventricular failure, including eight patients without peripheral edema and seven patients with massive peripheral edema. Baseline hemodynamic parameters were similar in both groups. In the patients without edema sublingual nitroglycerin decreased the left ventricular filling pressure from 28 ± 2 to 14 ± 3 mmHg and increased cardiac output from 3.3. ± 0.4 to 4.3 ± 0.3 L/min. Right atrial pressure decreased from 28 ± 3 to 13 ± 3 mmHg. In the group with massive peripheral edema no significant hemodynamic response to nitroglycerin was observed. Five patients from this group were rechallenged with sublingual nitroglycerin when peripheral edema had cleared after 18 days of furosemide therapy. At this time there were significant decreases in left ventricular filling pressure (from 13 ± 2 to 9 ± 2 mmHg) and right atrial pressure. Cardiac output was not altered. These findings suggest that peripheral edema prevents nitroglycerin from reducing afterload. However, they do not indicate that nitrates should be given after maximal effect of diuretics has been achieved, as the use of diuretics alone almost normalizes (or occasionally impairs) the hemodynamic parameters. It can only be concluded that some patients may benefit from diuretics only.

Armstrong et al [171] reported a pharmacokinetic-hemodynamic study supporting the view that systemic congestion may result in resistance to nitrates in congestive heart failure. They found that patients with systemic congestion and right atrial pressures of 19 ± 4 mmHg responded poorly to intravenous nitroglycerin, although they had higher plasma levels of the drug than patients with right atrial pressures of 10 ± 4 mmHg who responded favorably.

In summary, nitrates produce a rapid, and usually also sustained, hemodynamic improvement in patients with congestive heart failure. The primary effect is a reduction in left ventricular filling pressure. This is associated with clinical improvement – mainly relief of congestive symptoms but also an improvement in exercise capacity.

10.2. In What Stage of Heart Failure Should Nitrates Be Given?

Until now all studies of nitrates in congestive heart failure have been performed in moderate to severely symptomatic patients with pump failure. It may be claimed that the logic of use for nitrates applies also to mildly symptomatic or asymptomatic patients with pump failure. Moreover, nitrates might theoretically prevent progress of the disease if initiated at early stages. However, as no studies have confirmed this hypothesis, such treatment should be regarded at present as experimental.

Nitrates are clearly indicated in patients with functional class II–IV. In acute-onset congestive heart failure due to acute myocardial infarction nitrates may be used very early, when the first clinical symptoms and/or signs of congestive heart failure develop. Many authorities recommend the use of nitrates even before development of such symptoms or signs if hemodynamic impairment is observed in patients with a large infarction.

10.3. Do Nitrates Have a Direct, Positive Inotropic Effect?

Nitrates are usually considered to be vasodilators without any direct effect on myocardial contractility. However, the hemodynamic improvement produced by nitrates in congestive heart failure has stimulated investigation of the potential myocardial effects of these agents.

Several recent studies reveal that nitrates may improve myocardial contractility, perhaps by an improvement in the myocardial oxygen supply/demand ratio [172, 173]. Recently Bonoron-Adèle et al [174] studied the effect of nitroglycerin and nitroprusside on contractility and relaxation of electrically stimulated cat papillary muscles during hypoxia and reoxygenation. A positive inotropic effect was observed at the last phase of recovery. It was accompanied by an improvement in relaxation. This effect was evident regardless of the condition of the muscle and the moment at which nitrates diffuse into the muscle.

Calcium antagonists abolished the positive inotropic effect of nitrates. This suggests that the inotropic effect is related to modification of the effect of hypoxia on the calcium current.

It should be mentioned that several investigators have recently claimed that some "direct-acting" vasodilators other than nitrates exert a positive inotropic effect. For example, Khatri et al [175] reported that intracoronary administration of hydralazine, 0.8 mg, increased myocardial contractility in the perfused zone. Leier et al [176] reported that hydralazine exerts a positive inotropic effect in patients with left ventricular dysfunction and congestive heart failure. In the in vivo studies the augmentation of myocardial contractility may result from increased sympathetic activity due to vasodilation. However, the studies of isolated myocardial preparations and of intracoronary administration of the vasodilators indicate that nitrates and hydralazine might have a direct positive inotropic effect.

The discussion as to whether nitrates have a positive inotropic effect is of no clinical significance by itself, because even if such effect does exist, it is probably not very potent. The main importance of this effect is when nitrates are compared with other vasodilators, such as calcium antagonists which have a negative inotropic effect, in congestive heart failure. In this condition the vasodilators with positive (even if small) or no inotropic effect are theoretically superior to those with a negative inotropic effect.

10.4. Effect of Nitrates on the Atrial Natriuretic Peptide in Congestive Heart Failure

The atrial natriuretic peptide is a hormone secreted from cardiac tissues (atria as well as ventricles) in response to elevated pressure, stretch and other stimuli. It causes natriuresis and diuresis and is participating in fluid volume regulation. Recently Obata et al [176a] have shown that plasma atrial natriuretic peptide levels decreased after infusion of ISDN in patients with chronic congestive heart failure. These levels had a significant positive correlation with preload and negative correlations with plasma aldosterone and angiotensin II levels. This response to ISDN can be explained by reduced preload and atrial stretch, resulting in a reduction in the hormone secretion.

11. Congestive Heart Failure and Valvular Diseases

11.1. Chronic Valvular Diseases

While it is appreciated that vasodilators are effective in patients with congestive heart failure, their use in those patients in whom heart failure results from valvular diseases (particularly aortic and mitral stenosis) has been questioned by some investigators. Recent studies have revealed, however, that nitrates are useful in some of these patients.

Rabinowitz et al [156] studied three patients with congestive heart failure treated with intravenous ISDN. Two of them with rheumatic valvular insufficiency improved and were sent to surgery while receiving the drug. The third patient, with combined aortic and mitral insufficiency, required combined treatment of ISDN and dopamine. These investigators also reported that intravenous ISDN produced a hemodynamic response with a shift of the left ventricular function curve upward and to the left in patients with valvular insufficiency and elevated systemic vascular resistance treated with ISDN. Nitroglycerin was also found effective in aortic regurgitation. It was shown to decrease left ventricular filling pressure and to relieve symptoms in patients with this valvar lesion. Although systemic vascular resistance was also reduced, the decrease in left ventricular filling pressure precluded a significant increase in cardiac output. It was also suggested that by reducing left ventricular wall stress chronic treatment with nitrates may delay the progressive development of irreversible myocardial damage in aortic regurgitation. This hypothesis is yet to be confirmed.

Klein et al [178] studied the effects of intravenously administered nitroglycerin, given for one to several days, in 12 patients with mitral and/or aortic valvular defects and advanced refractory congestive heart failure (New York Heart Association class IV). The acute administration of nitroglycerin led to significant decreases in right atrial pressure, pulmonary artery pressure and mean pulmonary capillary wedge pressure. This response was maintained throughout the continuous nitroglycerin infusion. Mean systemic blood pressure and systemic vascular resistance were not significantly altered. Cardiac output was not altered in the whole group, but there was, however, a markedly divergent behavior of individual values. No correlation was found between the effect of nitroglycerin on cardiac output and type of the valvar lesion.

Goppel et al [179] studied the effects of ISDN on left ventricular contractility, volume and wall stress in ten patients with chronic aortic regurgitation (regurgitation fraction greater than 45%). The drug reduced preload and, to a slight extent, also afterload. Modest increases were observed in parameters of isovolumetric contractility due to the reduction in preload. Parameters of ejection phase contractility improved due to afterload reduction. Left ventricular ejection fraction increased from 57% ± 5% to 63% ± 7%. Total and effective stroke volume were reduced and the regurgitation fraction was not altered. It was concluded that ISDN may be used to reduce left ventricular end-diastolic pressure in patients with aortic regurgitation, but this may be at the expense of a reduction in left ventricular pump function.

Somewhat different results were obtained by Baxley et al [180], who studied the acute hemodynamic effects of sublingual nitroglycerin in ten patients with valvar lesions, as part of a presurgical evaluation. Four had aortic regurgitation, three had mitral regurgitation and three had both. The primary effect of the drug was a decrease in preload with diminished end-diastolic pressure, volume and wall stress. There was usually also a slight decrease in systemic vascular resistance, systolic ventricular pressure and end-systolic volume. Systolic wall stress was generally decreased. Ejection fraction increased, regurgitant flow and total stroke volume decreased, and forward cardiac output was slightly augmented.

Decrease of left ventricular volume and mitral regurgitant orifice has been suggested as a potential mechanism of reduction of mitral regurgitation. Nitrates may exert this effect by venodilation and preload reduction.

Roth et al [181] studied the effect of intravenous nitroglycerin on left ventricular hemodynamics, dimensions and function in 13 patients with chronic severe mitral regurgitation. The drug significantly reduced heart rate from 95 ± 20 to 89 ± 18 beats/min, mean blood pressure from 92 ± 11 to 79 ± 12 mmHg, right atrial pressure from 12 ± 5 to 7 ± 4 mmHg, pulmonary capillary wedge pressure from 23 ± 6 to 14 ± 6 mmHg, systemic vascular resistance from 1986 ± 524 to 1576 ± 557 dyn sec cm^{-5}, end-diastolic left ventricular volume from 248 ± 54 to 214 ± 53 ml, and end-systolic volume from 132 ± 61 to 97 ± 44 ml. Cardiac index, regurgitant fraction and regurgitant volume were not significantly altered. It was concluded that the beneficial effects of nitroglycerin on left ventricular function resulted from peripheral vasodilation and were not due to improvement of mitral regurgitation.

Valvular stenosis should be evaluated independently. Kawashita et al [182] reported that in ten patients with mitral stenosis and pulmonary congestion ISDN reduced pulmonary arterial pressure and cardiac volumes. Combination with dobutamine also resulted in an increased cardiac index.

In severe aortic stenosis vasodilators should be used with great caution, if at all. In my experience sublingual nitroglycerin relieved symptoms in an elderly patient with moderate aortic stenosis and pulmonary edema. In another patient ISDN oral spray relieved symptoms of prepulmonary edema. In aortic stenosis an additional problem of anginal pain may be present. Sublingual nitroglycerin was shown to reduce left ventricular filling pressure and relieve chest pain without critically reducing systemic arterial pressure or significantly altering cardiac output [183]. Nitroglycerin may be used to treat anginal symptoms in patients with aortic stenosis, with or without coronary artery disease.

All the studies mentioned above evaluated the acute effects of nitrates in valvar diseases. Somewhat different results were obtained in long-term studies of patients with aortic stenosis. Tebbe et al [184] performed invasive hemodynamic studies in nine patients with severe aortic stenosis (mean aortic valve gradient 90 ± 25 mmHg) before and after 22 ± 11 days of treatment with ISDN, 200 mg/day orally. At comparable exercise levels mean pulmonary artery pressure did not change significantly. Left ventricular ejection fraction increased slightly but not significantly from 66% ± 14% to 69% ± 9%. Only two patients with significantly depressed left ventricular resting function showed a distinct increase in ejection fraction after ISDN. Thus, in severe aortic stenosis the effects of chronic treatment with ISDN do not correspond to the beneficial effects of acute administration of nitrates. The results maybe, however, also due to tolerance to the high dose used.

In summary, nitrates may be tried in patients with acute or chronic valvular regurgitation and congestive heart failure. Nitrates are probably most effective in acute aortic regurgitation, as well as in acute mitral regurgitation complicating myocardial infarction. They may produce short-term relief in cases of aortic stenosis.

11.2. Acute Valvular Diseases

The experience with nitrates in acute valvular diseases is limited to a few experimental studies and clinical reports.

Klepzig et al [184a] produced ventricular volume overload in 18 dogs by constructing a valved conduit from the descending thoracic aorta to the left ventricular apex. Nitroglycerin, titrated to reduce mean aortic blood pressure by 40%, and placebo, were infused for 20 min each. Compared with placebo, nitroglycerin significantly reduced aortic flow, regurgitant flow, left ventricular end-diastolic diameter and pressure, left ventricular stroke work and systemic vascular resistance. Pulmonary flow, coronary flow and subendocarcial pH were unaltered.

These findings show that nitroglycerin can reduce afterload and preload while maintaining myocardial oxygen supply in the setting of acute volume overload. As acute volume overload is the hemodynamic impairment in acute aortic regurgitation, these findings support the use of nitrates in this condition.

Two clinical reports support the use of nitrates in acute aortic regurgitation. Sublingual ISDN was shown to increase cardiac output and reduce the left ventricular filling pressure in a patients with acute-onset heart failure due to acute aortic regurgitation caused by bacterial endocarditis [184b]. In another report nitroprusside was shown to decrease the regurgitant flow and the left ventricular end-diastolic pressure and to increase cardiac index and ejection fraction in patients with acute aortic regurgitation [184c]. These reports were published 13 years ago. Since than no clinical experience with nitrates in acute aortic regurgitation has been reported.

12. Long-Term Nitrate Therapy in Congestive Heart Failure

Long-term use of nitrates in congestive heart failure depends on persistence of the effect throughout the period of treatment. This is valid for both prolonged intravenous administration (more than a few days) and chronic oral or transdermal therapy.

Most clinical studies on long-term treatment and the potential development of tolerance to nitrates have been performed with ISDN, as this was the first widely used nitrate formulation to achieve relatively constant plasma concentration of the drug.

Several studies have revealed that the hemodynamic effects of nitrates in congestive heart failure are sustained throughout prolonged treatment. For example, Franciosa and Cohn [90] compared the effects of ISDN, 160 mg daily, and placebo in patients with chronic congestive heart failure over 3 months. The initial decrease in pulmonary capillary wedge pressure induced by the drug was sustained in repeated hemodynamic evaluation after 3 months. No such effect was observed in the group receiving placebo. Other clinical studies also revealed that long-term administration of ISDN to patients with chronic congestive heart failure is associated with sustained efficacy [185, 186].

Natarajan et al [186a] have recently shown that the hemodynamic effect of intravenous nitroglycerin is sustained for 48 h. They performed hemodynamic measurements in 15 patients with chronic congestive heart failure before and during intravenous nitroglycerin therapy. The drug was initiated at an infusion rate of 10 μ/min, with 5-minute increments until a hemodynamic response appeared with a decline of at least 5 mmHg in the right or left ventricular filling pressures or a 20% decrease in systemic vascular resistance. Once a significant hemodynamic effect was achieved, the nitroglycerin infusion was maintained at this rate for 48 h. A significant and sustained hemodynamic improvement was observed, without evidence of tolerance.

These results contrast with the findings of Packer et al [186b], who observed early tolerance in patients with heart failure during continuous infusion of nitroglycerin. Packer et al [186b] have used, however, only high constant infusion rates, and not low doses titrated to achieve a therapeutic response, as used by Natarajan et al [186a].

The sustained hemodynamic improvement is accompanied by clinical improvement. It is my experience (from a series of hundreds of patients with congestive heart failure) that relief of congestive symptoms is sustained for long periods of treatment. Leier and coworkers [92] evaluated this aspect in a double-blind, placebo-controlled trial. They revealed symptomatic improvement and an increase in exercise tolerance in patients with chronic congestive heart failure treated for 3 months with ISDN. These findings indicate sustained efficacy of nitrates without development of tolerance.

Prolonged efficacy was observed also with mononitrates. Klütsch et al [187] reported noninvasive evaluation of IS-5-MN in patients with chronic congestive heart failure. After prolonged treatment the cardiothoracic index was significantly reduced in all patients. Other indices also revealed improvement.

Other investigators suggested that at least partial tolerance to the circulatory effects develops rapidly during sustained therapy with ISDN. One of the major problems in evaluating this suggestion is that not only is the number of controlled studies limited, but tolerance in each indication (such as angina pectoris, congestive heart failure, etc.) should be independently evaluated.

The opinion that partial tolerance does develop depends on mixed findings from various conditions:

1. The improvement of exercise tolerance after any dose of ISDN was found to be greater during acute than during sustained therapy. The duration of improvement in exercise capacity was also found to be greater after a single dose than during sustained therapy [108].
2. In normal volunteers and in patients with chronic congestive heart failure the decrease in systolic arterial pressure and increase in heart rate for any given dose of nitrate were smaller and lasted for a shorter duration during prolonged than during acute therapy [188, 101].
3. Animal experiments revealed that for up to 1 week of repeated nitroglycerin administration the doses required to achieve a given decrease in blood pressure were up to 50 times higher than the doses required at initiation of the study [189, 190].
4. The venodilator effect of nitrates, probably their main beneficial effect in congestive heart failure, might be attenuated during prolonged therapy [154, 191].
5. The headache accompanying the initiation of nitrate therapy disappears within a few days.
6. Industrial workers exposed to nitrates have developed withdrawal symptoms, suggesting nitrate dependence. Dependence is usually associated with tolerance. These workers did not, of course, have congestive heart failure.
7. The response to acute intravenous administration of ISDN was attenuated in patients treated for long periods with ISDN ointment [192].
8. Clinical improvement may be attenuated despite sustained hemodynamic improvement. This was the case in patients with chronic congestive heart failure, treated for 3 months, as reported by Franciosa and Cohn [90].

Despite these findings most investigators are convinced that there is no significant tolerance, and that data suggesting such a tolerance are insufficient. Reviewing the subject in 1980 Abrams [193] stated that "there is little evidence that chronic administration of organic nitrates even in large doses, induces clinically important tolerance or cross-tolerance to nitroglycerin or other nitrates in patients being treated for angina pectoris or heart failure". I completely agree with this statement.

The question of whether tolerance to transdermal administration of nitrates (by discs) develops is discussed in the chapter on transdermal nitrates. The conclusion is that tolerance to nitrates probably does not develop if a drug-free interval is allowed, and the efficacy is sustained.

It should be emphasized that even those who believe that there is some tolerance to nitrates admit that most problems concerning tolerance have been reported in patients with angina pectoris rather than in those with heart failure.

Similarly to the case in angina pectoris, potential development of tolerance to nitrates in congestive heart failure may be prevented by intervals during which nitrates are not given, or the plasma level of nitrates is markedly reduced. To assess whether effectiveness can be maintained by intermittent use Hall et al [194] measured systemic blood pressure, right atrial pressure, pulmonary artery pressure, heart rate, cardiac output, ejection fraction and plasma concentration of ISDN before and for 24 h after a single, morning, oral dose of ISDN, sustained release, 120 mg, on the 1st and 7th days of treatment in 13 patients with chronic congestive heart failure. Blood pressure, cardiac output, heart rate, and ejection fraction were unchanged, while right atrial pressure decreased. Mean pulmonary arterial pressure was lowered from 33.6 to 20.5 mmHg on day 1 and from 30.4 to 22.0 mmHg on day 7. The reduction was maximal at 1–3 hours, persisted at significant levels for 9–10 hours and then gradually returned to baseline. During exercise mean pulmonary arterial pressure was reduced by 19% on days 1 and 7. Plasma concentration of the drug was maximal at 1–2 h after administration and fell to near baseline levels at 24 hours on days 1 and 7. Hence, once daily administration of slow-release ISDN, 120 mg, produced plasma levels that were sufficient to reduce left ventricular filling pressure throughout daytime in the period of maximal physical activity. The intervals of low plasma levels during the night prevent the development of tolerance.

Makhoul et al [194a] have recently shown differential venous pulmonary and systemic effects of continuous nitrate therapy in patients with heart failure. After 24 h of intravenous nitroglycerin infusion right atrial pressure returned to pretreatment level while the beneficial effect on systemic vascular resistance, cardiac index and pulmonary capillary wedge pressure was largely maintained.

Packer et al [194b] studied the role of neurohormonal activation in the development of nitrate tolerance in patients with chronic congestive heart failure. Fifty two patients received various dosing regimens of nitroglycerin intravenously: a continuous infusion or drug free intervals of 4, 8 and 12 h. The continuous and 4-hours interval infusion produced the greatest increases in plasma renin activity and heart rate, although the 12-hours interval infusion produced the greatest decrease in mean arterial pressure (and thus should have theoretically been associated with the greatest increases in plasma renin activity and heart rate). The continuous infusion was associated with more frequent development of tolerance than the other regimens. These findings indicate that neurohormonal activation may contribute to the development of tolerance to nitroglycerin in patients with heart failure.

12.1. Effects of Nitrates on Mortality in Congestive Heart Failure

For many years nitrates were used in patients with chronic congestive heart failure without knowing their effect on mortality. Only recently have the results of the V-HeFt study been published, showing a reduction in mortality by combined therapy with ISDN and hydralazine. Francis et al [195] studied 186 patients with heart failure who were randomized to treatment with ISDN and hydralazine and 273 patients randomized to placebo. Long-term mortality was reduced by 28% in the patients treated with the vasodilators. In patients with coronary artery disease mortality was reduced by 28.5% and in those with noncoronary disease by 27.4%. In those with hypertensive history mortality was reduced by 38.4% and in those without such history by 23.3%; in those with alcoholic history by 20.9% and with nonalcoholic history by 36.1%. Antiarrhythmic drug treatment history was associated with a reduction of 31.1% and nonantiarrhythmic history with one of 24.2%. In patients with ejection fraction less than 28% mortality was reduced by 35.5%, while in those with ejection fraction over 28% it was reduced by 20.7%. In patients with oxygen consumption during exercise of less than 14.5 ml/kg per minute (median value) mortality was reduced by 23.0%, while in those with oxygen consumption above this level mortality was reduced by 39.2%. The differences of reduction in mortality between the subgroups did not reach statistical significance.

In conclusion the combination of ISDN and hydralazine appears to reduce mortality in patients with congestive heart failure regardless of etiology or clinical severity of heart failure [195]. With this study ISDN became the first drug shown to reduce mortality in chronic congestive heart failure.

In the following years further data on analysis of subgroups in the V-HeFT population have been published. Cohn and his coworkers [195a] have continued the trial and included 642 patients followed for a mean period of 2.3 years. They found that baseline variables do not predict sudden or pump failure death, and effective vasodilator treatment with the combination of ISDN and hydralazine reduces carciac mortality of all causes. Moreover, these investigators [195b] reported that the benefit of ISDN plus hydralazine as compared to placebo and prazosin appears to be confined to patients with ventricular arrhythmias at all levels of ventricular dysfunction.

These findings are in contrast to those of the CONSENSUS trial which showed that the angiotensin converting enzyme enalapril reduces mortality when added to conventional therapy in patients with severe heart failure. In these trial only mortality related to progresive pump failure was reduced, while sudden arrhythmic death was unaltered.

The V-HeFT 2 study, comparing the two types of vasodilator therapy is now ongoing. Its target is to reveal whether the combination of ISDN and hydralazine, or an angiotensin converting enzyme inhibitor, are superior in reducing mortality in chronic congestive heart failure.

The VHeFT findings are of particular importance in view of a tendency towards some disappointment from the achievements of antiarrhythmic drugs and appreciation of their potential risk, mainly in patients with ventricular dysfunction.

13. Nitrates Combined and Compared with Other Drugs in Congestive Heart Failure

13.1. Nitrates in Combined Therapy for Congestive Heart Failure

In patients with congestive heart failure the efficacy of nitrates may be enhanced by combination with other drugs. Nitrates are mainly venodilators. Therefore it is possible that combination with potent arteriolar dilators and/or positive inotropic agents may widen the therapeutic spectrum and, in particular, achieve a greater increase in cardiac output than that produced by nitrates alone.

13.1.1. Hydralazine

Hydralazine is a potent arteriolar dilator which reduces the systemic vascular resistance without directly affecting the venous circulation. It may reduce pulmonary capillary wedge pressure by improving left ventricular empting, despite the lack of venodilatory effect. However, at least 40% of patients with congestive heart failure treated with hydralazine also require nitrates to reduce the pulmonary capillary wedge pressure [196].

A large Veterans Administration cooperative study (V-HeFt) was performed to evaluate, among other things, the efficacy of combined therapy with nitrates and hydralazine in congestive heart failure. The arteriolodilatory effect of hydralazine combined with the venodilatory effect of nitrates should, theoretically, improve the hemodynamic response.

By lowering systemic arterial pressure and, as a reflexion response, accelerating heart rate, hydralazine may aggravate myocardial ischemia. The incidence of hydralazine-induced ischemic events was found to be higher in patients on hydralazine monotherapy than in those concomitantly treated with nitrates. Moreover, hydralazine-induced deterioration in exercise performance in some patients with congestive heart failure and coronary artery disease was reversed by nitrates but not by propranolol [197]. The finding has stimulated extensive research into the combination of hydralazine and nitrates in congestive heart failure.

In intensive care units it is common practice to start treatment in patients with congestive heart failure with nitrates as the only vasodilator. If response is inadequate, particularly the decrease in systemic vascular resistance and increase in cardiac output, hydralazine is added. Several investigators found that this combined treatment produced greater hemodynamic improvement than either drug alone.

Massie et al [198] reported that in 12 patients with chronic congestive heart failure hydralazine lowered systemic vascular resistance and increased cardiac output without altering left ventricular filling pressure. In contrast, nitrates decreased left ventricular filling pressure without altering the other parameters. Combined therapy resulted in improvement in all hemodynamic parameters: a mean decrease of 36% in left ventricular filling pressure and a mean increase of 58% in cardiac index. Hemodynamic improvement was observed at low work load but not at maximal exercise. The hemodynamic response was associated with sustained clinical improvement in seven of the patients who received long-term combined therapy.

Because both agents are potent vasodilators, postural hypotension, tachycardia and even myocardial ischemia may complicate their combined use [199, 200]. It is important to perform upright and supine hemodynamic evaluation at the initiation of this combined therapy.

The effects of ISDN and hydralazine on mortality in patients with congestive heart failure are discussed in another section of this chapter.

13.1.2. Captopril

Captopril and nitrates have a comparable but not similar hemodynamic profile. Their combined administration requires caution to avoid excessive reduction of left ventricular filling pressure or systemic arterial pressure. Nevertheless, some patients with severe congestive heart failure may respond only to the combination of these agents. In my experience several patients have been effectively and safely treated with this combination.

Sublingual nitroglycerin tablets, intravenous nitroglycerin or ISDN oral spray can be effective in patients with congestive heart failure who develop acute hemodynamic deterioration and pulmonary edema during long-term treatment with captopril.

Lewis et al [201] studied the effects of captopril and ISDN in 11 patients with severe heart failure. Captopril increased cardiac, stroke and stroke work indices while reducing systemic impedance, heart rate and double product. Contractility was unchanged or decreased slightly, following a direct or reflex sympatholytic effect. Sublingual administration of ISDN at individually titrated doses enhanced the hemodynamic effect of captopril and, in addition, lowered filling pressures. Addition of dobutamine clearly increased the inotropic state of the left ventricle and produced a further increase in cardiac index.

Captopril can block the activation of the renin-angiotensin system by nitrates.

Recently Haude et al [201a] compared the acute hemodynamic effect of sublingual captopril versus nitroglycerin in patients with heart failure. They found a somewhat greater effect of captopril, but as no dose titration was made the difference can be attributed to uncomparable doses. The time to peak effect is not given in their publication but undoubtly nitroglycerin acted more rapidly. The effect of captopril was sustained for a longer period, but this is not surprising in view of the different half-lives of both drugs.

The major problem in using captopril acutely, by sublingual administration for treatment of acute deterioration of heart failure is safety. Patients with heart failure are usually on diuretics and may be hyponatremic. Such patients are at risk of first dose hypotension after administration of captopril or other angiotensin converting enzyme inhibitors.

In our view there is no use to replace sublingual nitrates by sublingual captopril for treatment of acute-onset heart failure or acute deterioration of chronic heart failure.

13.1.3. Sympathomimetic Amines

Nitrates may, particularly at high doses, produce an excessive reduction in systemic arterial pressure. In some cases of congestive heart failure the beneficial effect of reduction in left ventricular filling pressure cannot be achieved without such an excessive reduction in systemic pressure. In other cases the vasodilator effect of nitrates cannot be fully expressed as an increase in cardiac output due to impaired myocardial contractility. The combination with sympathomimetic amines such as dopamine and dobutamine can augment myocardial contractility and prevent a decrease in arterial pressure. I have found this combination particularly useful in patients with ischemic heart disease, pulmonary edema and hypotension. There is good, although limited, experience in the combined use of ISDN with both dobutamine and dopamine.

The combination of nitrates and sympathomimetic amines may be particularly useful in patients with cardiogenic shock. Bussmann et al [202] studied the hemodynamic response to dobutamine, alone or in combination with various high doses of nitroglycerin, in 12 patients with cardiogenic shock. In seven patients with severe valvar disease dobutamine significantly increased cardiac output from 4.0 to 4.5 L/min without changes in systemic arterial pressure and left and right ventricular filling pressures. Additional intravenous infusion of nitroglycerin (1.5–3.0 mg/h) increased cardiac output to 4.7 L/min and reduced filling pressures. Systemic arterial pressure remained stabilized. Higher doses of nitroglycerin reduced the filling pressures considerably, but cardiac output and systemic arterial pressure fell.

In five patients with cardiogenic shock due to severe myocardial damage nitroglycerin did not enhance the increase in cardiac output produced by dobutamine. At high doses of nitroglycerin filling pressures fell, but so did also systemic arterial pressure.

It may be concluded that the treatment of cardiogenic shock by a combination of dobutamine (7 μg/kg per minute) and nitroglycerin (1.5–3.0 mg/h) produced hemodynamic improvement in some patients.

It should be remembered that the optimal left ventricular filling pressure is higher in patients with cardiogenic shock than in those with milder forms of heart failure. This should guide the dose titration of nitrates in cardiogenic shock.

13.1.4. Digitalis Glycosides

It is common practice to add nitrates to conventional treatment with digitalis and diuretics in patients with congestive heart failure whose symptoms are not adequately controlled. Orally administered ISDN or transdermally administered nitroglycerin may be particularly effective in patients who suffer mainly from congestive symptoms rather than symptoms of low cardiac output. Sublingual administration of nitroglycerin, ISDN tablets or ISDN oral spray produce hemodynamic and symptomatic improvement in patients who develop acute deterioration of congestive heart failure and pulmonary edema while on long-term treatment with digitalis and/or diuretics.

13.1.5. *N*-Acetylcysteine

N-acetylcysteine may reverse tolerance to nitrates and enhance their effect. Bertel and Noll [203] reported that nitroglycerin responsiveness is enhanced by *N*-acetylcysteine infusion in patients with congestive heart failure, especially in those with a blunted responsiveness. Therefore, the combination may be beneficial during long-term nitrate therapy.

13.2. Nitrates Compared with Other Drugs in Congestive Heart Failure

13.2.1. Hydralazine

Comparative studies of nitrates and hydralazine have been important not only for specific evaluation of these agents but also for understanding the pathophysiology of congestive heart failure and the role of selective vasodilation. The main difference between these agents is their vascular selectivity. Hydralazine is an arteriolar dilator whereas nitrates dilate mainly the venous vasculature (particularly at low doses).

Massie et al [204] found that in patients with congestive heart failure hydralazine augmented cardiac output and renal and limb blood flow. ISDN increased cardiac output to a lesser extent and did not alter renal or limb blood flow. Neither drug, nor their combination, changed hepatic blood flow. ISDN reduced pulmonary capillary wedge pressure more than hydralazine did.

At present venodilation is considered the predominant therapeutic effect of vasodilators in congestive heart failure. Therefore nitrates are superior to hydralazine. Moreover, it is not recommended to give hydralazine in acute-onset ischemic pump failure, a condition in which nitrates are beneficial. Another problem with hydralazine is the rapid tolerance which many patients with chronic congestive heart failure develop to it.

13.2.2. Captopril

Captopril is an angiotensin-converting enzyme inhibitor with balanced vasodilatory properties. Comparative studies have revealed that captopril and ISDN can produce similar decreases in left ventricular filling pressure and systemic vascular resistance. However, heart rate may be slowed by captopril due to a negative chronotropic effect, whereas nitrates often accelerate heart rate. Therefore the vasodilatory effect of nitrates may be expressed as a greater increase in cardiac output than that of captopril. This hypothesis however, is yet to be confirmed.

Another important difference is the time course of response. Patients may show either no response, a sustained response or an initial response which is later attenuated, to nitrates as well as to captopril. However, about 15% of patients treated with captopril but not with nitrates show another pattern, a triphasic response. These show an initial improvement, followed over subsequent days by attenuation and, finally, late recurrence of the initial effect [205].

Unlike nitrates, captopril is not rapidly effective in patients with acute deterioration of congestive heart failure and pulmonary edema because it can be given only orally. Enalaprilat, a metabolite of the angiotensin-converting enzyme enalapril, may be given intravenously, but it has not been studied in these conditions.

Packer et al [206] reported a comparative study of captopril, 25 mg orally, and ISDN, 40 mg orally, in 18 patients with severe chronic congestive heart failure. The randomized crossover study was conducted on consecutive days. Captopril and ISDN produced similar decreases in systemic vascular resistance. However, ISDN reduced pulmonary vascular resistance significantly by 25%, and captopril reduced it nonsignificantly, by only 5%. The difference between the two drugs was highly significant. Left ventricular filling pressure declined similarly with both captopril and ISDN, but because pulmonary arteriolar resistance fell significantly with ISDN, mean right atrial pressure decreased more with ISDN (5.4 mmHg) than with captopril (2.8 mmHg). Systemic vascular resistance declined similarly with both drugs. Cardiac index increased more with ISDN (0.47 L min^{-1} m^{-2}) than with captopril (0.23 L min^{-1} m^{-2}), and therefore mean arterial pressure declined less with ISDN (10.5 mmHg) than with captopril (16.7 mmHg). The difference in the effects of these drugs on cardiac index did not result from differences in their effects on heart rate, since heart rate was similarly slowed with both drugs, and thus both drugs similarly increased the stroke volume index.

These data indicate that in patients with severe chronic congestive heart failure nitrates produce pulmonary arteriolar vasodilation not shared by captopril. As captopril does not reduce right ventricular afterload, the extent to which it can increase the right ventricular output is limited. This may explain the greater hypotensive effect of captopril, despite similar effects of both drugs on left ventricular filling pressure and systemic vascular resistance.

This difference may result in different clinical effects of both drugs. In the study by Packer et al [206] two patients developed symptomatic hypotension with captopril whereas none did so with ISDN.

13.2.3. Diuretics

Despite the frequency with which diuretics and nitrates are used in chronic congestive heart failure, no prospective, randomized, comparative study of their hemodynamic effects in this condition has been reported. Only recently a comparative study of these two commonly used drugs in acute heart failure complicating myocardial infarction has been reported. Nelson et al [207] compared the acute hemodynamic effects of intravenous furosemide, 1 mg/kg, and intravenous ISDN, 50–200 μg/kg per hour, in 28 patients with left ventricular failure complicating acute myocardial infarction. Furosemide reduced the left ventricular filling pressure and cardiac output and transiently elevated systemic blood pressure. ISDN reduced systemic blood pressure, systemic vascular resistance, pulmonary arterial pressure and left ventricular filling pressure without altering the cardiac output. ISDN was therefore superior to furosemide, as it did not impair peripheral perfusion and reduced systemic vascular resistance.

Hutton et al [208] compared the hemodynamic effects of furosemide, 0.05 mg/kg intravenously, with those of IS-5-MN, 10–20 mg intravenously, in ten patients with chronic heart failure and left ventricular dysfunction. Hemodynamic measurements were performed via Swan-Ganz and intra-aortic catheters. Coronary flow was measured by a thermodilution catheter in the coronary sinus. In the patients treated with IS-5-MN heart rate increased from 79 ± 6 to 87 ± 9 beats/min, systolic blood pressure fell from 126 ± 18 to 119 ± 17 mmHg, and mean blood pressure fell from 98 ± 10 to 95 ± 12 mmHg. Pulmonary capillary wedge pressure fell from 15 ± 1 to 6 ± 2 mmHg with an increase in cardiac output from 4.6 ± 0.4 to 5.0 ± 0.4 L/min. Coronary venous flow increased from 124 ± 20 to 157 ± 39 ml/min and AV oxygen difference narrowed from 11.8 ± 4.0 to 10.5 ± 0.7 vol%.

With furosemide heart rate was not altered but systolic blood pressure increased from 121 ± 11 to 142 ± 13 mmHg, and mean blood pressure increased from 96 ± 6 to 112 ± 9 mmHg. Pulmonary capillary wedge pressure increased from 17 ± 4 to 21 ± 5 mmHg but cardiac output was unchanged. Coronary blood flow increased from 125 ± 11 to 148 ± 30 ml/min, but AV oxygen was unchanged.

Although it may be agreed that many patients in this series did not have hemodynamic parameters compatible with heart failure, the results of this study emphasize the differences between hemodynamic effects of furosemide and nitrates. They confirm that intravenous loop diuretics produce acute vasoconstriction in patients with left ventricular dysfunction and suggest that nitrates may, due to their vasodilating activity, improve pump function in patients with heart failure.

These findings are not sufficient to make a general recommendation for first-line therapy in congestive heart failure. It should also be remembered that one of the mechanisms of the attenuation of long-term effect of nitrates is fluid retention, which may be overcome by diuretics. Moreover, in certain patients peripheral edema is the predominant sign of heart failure. At present it can only be said that in some patients nitrates may be superior to loop diuretics as first-line therapy. The regimen should, of course, be individually adjusted.

13.2.4. Summary - Nitrates in Congestive Heart Failure

It may be summarized that nitrates produce hemodynamic improvement in patients with acute and chronic congestive heart failure. This effect results mainly from venodilation and is associated with symptomatic improvement.

Nitrates are given mainly to patients with moderate to severe heart failure. The effect is usually sustained for long periods, but in some patients it may be attenuated during long-term therapy. The effects of nitrates are comparable or superior to those of other vasodilators. Nitrates are usually added to conventional treatment with digitalis and diuretics. In acute heart failure nitrates may be superior to diuretics. No comparative studies with digitalis have been reported. And, finally, nitrates can reduce mortality in chronic congestive heart failure.

More studies of the effect of nitrates on exercise tolerance in patients with heart failure are needed.

14. Systemic Hypertension, Arrhythmias and Cardiac Surgery

At moderate-to-high doses nitrates decrease systemic vascular resistance and may, thereby, lower elevated systemic arterial pressure. A decrease in left ventricular filling pressure and in cardiac output may also, in some cases without congestive heart failure, lower elevated arterial pressure. However, nitrates are not conventionally used for treatment of hypertension as their vascular and antihypertensive effect is rapidly attenuated during long-term oral therapy.

Nevertheless, the antihypertensive effect of nitrates is used in two conditions:

1. Hypertensive emergencies. The rapidity of onset of action and the fact that nitrates can be given rapidly as a sublingual tablet or oral spray make them suitable for treatment of hypertensive emergencies. Although no clinical studies have been reported, many centers have experience with this therapy in some patients. Nitrates may be particularly useful in patients who have also coronary artery disease.
2. Hypertensive episodes in the coronary care unit. Acute elevation of systemic arterial pressure often accompanies acute myocardial infarction, other ischemic episodes and cardiac surgery.

As safety of use for some new antihypertensive agents such as nifedipine in some of these conditions is yet to be determined, intravenously administered nitrates might be the optimal treatment. Nitrates are safe and might even reduce infarct size in patients with acute myocardial infarction. In the coronary care unit they should be given with hemodynamic monitoring. Appropriate reduction of systemic arterial pressure (not below levels impairing coronary perfusion) can contribute also to the anti-ischemic effect of nitrates. The widest experience in this condition is with intravenous ISDN. Some patients have been treated with IS-5-MN.

Caesar et al [209] studied the effects of ISDN, 20 mg orally, on blood pressure, peripheral vascular resistance and venous capacity in three groups: patients with arterial hypertension, patients with peripheral occlusive arterial disease and normal controls. Peripheral hemodynamics were determined by plethysmography. In all groups the drug lowered systolic and diastolic blood pressure.

The decrease was greatest in the hypertensive patients. All three groups showed an increase in arterial blood flow to the extremities and a decrease in peripheral vascular resistance. These changes were significantly more pronounced in the hypertensive patients and in the patients with peripheral vascular disease than in the healthy subjects. On the other hand, venous capacity increased to a comparable extent in all groups. It was also found that whole blood and plasma viscosity were significantly reduced following ISDN.

A most interesting finding has recently been reported by Duchier et al [210] who demonstrated that ISDN exerts a sustained antihypertensive effect in elderly patients with isolated systolic hypertension. They performed a double-blind, randomized study in 39 patients (mean age 80 ± 1 years) with isolated systolic hypertension in order to evaluate the antihypertensive effect of oral sustained-release ISDN, 40 mg twice daily, compared with placebo. After 12 weeks of treatment supine arterial pressure was reduced from 193 ± 10 to 166 ± 13 mmHg with ISDN ($p < 0.001$) and from 189 ± 10 to 176 ± 15 mmHg with placebo ($p < 0.001$). Analysis of variance revealed that the decrease in systolic blood pressure by ISDN (27 mmHg) was greater than that observed with placebo (13 mmHg). A similar response was observed for supine and standing systolic blood pressure measured at 8 and 12 h after administration of the drug. Heart rate and diastolic blood pressure were not significantly altered, and no significant adverse effects were observed.

At the end of the trial ISDN was tapered off gradually over 2 weeks. During this period systolic blood pressure gradually increased. It was not altered in the placebo group.

It may be concluded that in elderly patients with isolated systolic hypertension sustained-release ISDN induces a selective and sustained reduction in systolic blood pressure. The effect is observed up to 12 h after oral administration, and no tolerance is observed over long periods of treatment.

An important recent finding concerning the mechanism of action of nitrates is related to reflected pressure waves. Reflected pressure waves are an important contribution to the left ventricular hydraulic load. Nitrates were shown to reduce the systolic pressure load on the left ventricle by reducing the amplitude of reflected waves returning to the aorta [210a].

14.1. Nitrates in Ventricular Arrhythmias

Nitrates have no direct antiarrhythmic properties. As many cases of ventricular arrhythmias result from coronary artery disease, the anti-ischemic effect of nitrates may, however, exert an indirect antiarrhythmic effect.

Häge [211] compared the antiarrhythmic activity of IS-5-MN, 20 mg three times daily, and verapamil, 60 mg three times daily, in a double-blind, crossover, randomized study in ten patients with coronary artery disease accompanied by ventricular premature beats. Each drug was given for 2 weeks, after 1 week of placebo. All patients had, during the week on placebo, ventricular premature beats of Lown grade 2–4. With IS-5-MN ventricular premature beats were observed in three patients only, and they were of Lown grade 1. On verapamil five patients had Lown 2, and five had Lown 1 ventricular premature beats. IS-5-MN reduced exercise-induced ST-segment depression by 51%, from 2.1 to 1.0 mm, and verapamil by 15%, from 2.0 to 1.7 mm. IS-5-MN increased maximal work load, and decreased the frequency of anginal episodes more than verapamil.

This study represents an interesting new indication for nitrate therapy which should definitely be studied further. It has, however, some methodological problems regarding the counting of premature beats and the fact that the dose of verapamil was very low.

Senges et al [212] investigated the role of nitrates in arrhythmias after a chance clinical observation. A female patient with mitral valve prolapse and frequent ventricular premature beats developed chest pain and was given 0.8 mg nitroglycerin sublingually. Within a few minutes the arrhythmias had completely disappeared. This effect was reproducible in the subsequent clinical course.

Further studies by these investigators [212] revealed three patients with cardiac arrhythmias responsive to nitrates. The patients had mitral valve prolapse, severe ventricular arrhythmias and frequent familial occurrence of sudden death. Left ventricular function and coronary arteries were normal, and coronary reserve was not impaired. The ventricular arrhythmias were refractory to conventional antiarrhythmic therapy. However, they had almost completely disappeared within 1 min from administration of 0.8 mg nitroglycerin sublingually and within 30 min from oral administration of 20 mg sustained-release ISDN. A marked reduction of the arrhythmia was observed during treatment with ISDN, 20 mg four times daily.

Electrophysiologic studies have revealed that all three patients had normal intracardiac conduction and refractoriness. In one of the patients a reentry mechanism could be triggered by premature ventricular stimulation. Nitrates did not alter any electrophysiologic parameter except for the sinus rate, which was slightly accelerated. This acceleration probably resulted from sympathetic activation. Thus, nitrates had no effect on cardiac impulse propagation. Moreover, in vitro studies reveal that nitrates have no effect on the cell membrane of the normal or hypoxic conduction system.

In another study nitrates had a suppressant effect on supraventricular arrhythmias in postmyocardial infarction patients with elevated left ventricular filling pressure but not in those with normal filling pressure [213].

In one series of patients with ventricular premature beats and low left ventricular filling pressure nitroglycerin infusion resulted in increased number of the premature beats.

In summary, nitrates can exert an antiarrhythmic effect, despite the absence of any direct effect on impulse propagation, in three groups of patients:

1. Patients with coronary artery disease, in whom the anti-ischemic effect may result in the antiarrhythmic effect.
2. Some patients with mitral valve prolapse, in whom the mechanism of action is not clear.
3. Patients in whom the arrhythmias are due to elevated left ventricular end-diastolic pressure, which is reduced by nitrates.

Further studies in these three groups are definitely required.

14.2. Nitrates in Patients Undergoing Cardiac Surgery

Nitrates have been successfully used in management of perioperative hemodynamic disturbances, namely heart failure and systemic and pulmonary hypertension. Intraoperative and postoperative systemic arterial hypertension has been observed in about 60% of patients undergoing coronary artery bypass surgery [214, 215]. Risk factors for this complication are preoperative systemic hypertension, preoperative treatment with beta-adrenoreceptor blockers and left main coronary artery disease. Intravenously administered nitrates have been used in the past decade for treatment of this complication. Nitroglycerin lowered the elevated arterial pressure as well as the tension-time index and the magnitude of ST-segment depression.

Nitrates are effective also in combined perioperative hemodynamic disturbances. My colleagues have observed lowering of postoperative systemic and pulmonary hypertension in children undergoing surgery for coarctation of the aorta.

Kaplan and coworkers [216] studied the effects of intravenous nitroglycerin, given before and during coronary artery bypass surgery, in patients with preoperative systemic hypertension, elevated left ventricular filling pressure and ST-segment depression at rest. A mean dose of 110 ± 10 μg/min produced hemodynamic improvement (reduction in systemic pressure) in 87% of the patients. Higher doses were required during cardiopulmonary bypass and in patients with impaired left ventricular function.

It is important to evaluate the effect of nitrates in patients during anesthesia for noncardiovascular surgery to find out whether the hemodynamic profile is altered by the anesthesia itself. Zadrobilek et al [217] studied 17 patients in whom hypotension was induced by ISDN or nitroglycerin during neuroleptanesthesia for facial and neck surgery. The patients also received metoprolol and enflurane. During ISDN-induced hypotension mean arterial pressure was reduced from 83 to 63 mmHg and during nitroglycerin-induced hypotension from 81 to 53 mmHg. In both groups heart rate decreased by 10%. Cardiac index was not significantly altered while systemic vascular resistance had significantly decreased. Nitroglycerin, but not ISDN, reduced pulmonary arterial pressure and increased the intrapulmonary shunt volume. This hemodynamic profile is not different from that observed without anesthesia, however the anesthetic technique attenuated the activation of the adrenergic and renin-angiotensin system and suppressed reflex tachycardia which usually accompanies nitrate-induced hypotension.

15. Pulmonary, Abdominal, and Peripheral Vascular Effects

15.1. Effects on Pulmonary Vasculature

Theoretically nitrates might be useful in pulmonary hypertension due to noncardiac causes. However, only a small number of controlled studies have been reported, and their practical use is extremely limited.

Nitroglycerin, administered sublingually, was reported to decrease pulmonary vascular resistance and pulmonary arterial pressure. A pulmonary venodilatory effect is probably the main pulmonary effect of nitroglycerin. It contributes to the beneficial effects of the drug in congestive heart failure and ischemic heart disease [218].

Nitroglycerin can produce redistribution of pulmonary arterial perfusion in normal human subjects as well as in patients with cardiovascular or pulmonary diseases. This may result in impairment of the ventilation/perfusion ratio in the lungs and in an increase in pulmonary shunt flow. These effects may result from release of hypoxic vasoconstriction by the drug and may cause a reduction in arterial oxygen saturation [219]. This effect is smaller than that produced by nitroprusside and is of little clinical relevance.

Davis et al studied the effects of nitrates on pulmonary gas exchange in ten patients with normal cardiac and pulmonary function who received 0.6 mg nitroglycerin sublingually while breathing 100% oxygen. Blood gases and hemodynamics were measured before and 3 min after nitroglycerin. Administration of the drug was associated with a decrease in partial arterial oxygen tension from 324 ± 19 to 287 ± 24 mmHg and an increase in the alveolar-arterial oxygen tension from 356 ± 20 to 394 ± 25 mmHg. Since there was no change in the cardiac index and the mixed venous oxygen content, whereas the pulmonary vascular resistance decreased by 75 ± 26 dyn sec cm^{-5} m^{-2}, it appears that the nitroglycerin-induced hypoxemia is due to an increase in the intrapulmonary right-to-left shunt secondary to vasodilation of poorly ventilated pulmonary areas.

In patients with chronic obstructive pulmonary diseases and pulmonary hypertension nitroglycerin had both beneficial and potentially deleterious effects. It decreased the mean pulmonary arterial pressure, right atrial pressure, right ventricular end-diastolic volume index, cardiac index and arterial oxygen tension. It increased right ventricular ejection fraction and decreased or did not alter the total pulmonary vascular resistance [220, 221].

A combined study evaluating the effects of nitrates on both pulmonary vessels and airways is now being performed.

Ježek et al [222] studied the acute and chronic effects of ISDN in 18 patients with pulmonary hypertension due to chronic interstitial lung fibrosis. The pulmonary hypertension was refractory to acutely administered captopril and nifedipine. However, intravenous administration of ISDN resulted in a decrease of pulmonary arterial pressure by 35% ± 11% and pulmonary vascular resistance by 38% ± 13%. Sustained effect of oral ISDN was found in patients without progressive hypoxemia due to worsening of underlying disease. However, this therapy seemed not to be able to prevent a rapid rise of pulmonary hypertension and early death in patients with a malignant course of the disease. It was also found that the initial effect of ISDN is attenuated during chronic oral treatment. Moreover, the outcome of long-term treatment may not be reliably predicted from the results of first application.

15.2. Effect on Airways

Several investigators have reported that nitrates have a bronchodilating effect in asthmatics. Hirshleifer and Arora [223] reported that 0.3 mg nitroglycerin, sublingually, increased the FEV_1/FVC in patients with acute asthmatic episodes.

Okayama et al [224] reported that in 30 asthmatic patients 5 mg ISDN, sublingually, significantly reduced respiratory resistance and increased FEV_1 compared with placebo. ISDN tended to be less effective than metaproterenol, although in some patients the bronchodilatory effect was comparable to that of metaproterenol. Niehues el al [225] reported that nitroglycerin improved lung function and decreased airway resistance.

It is believed that the bronchodilatory effect results from relaxation of bronchial smooth muscle via nitrate receptors. Another possibility is that bronchodilation results from release of catecholamines due to nitrate-induced hypotension.

It should be emphasized that several investigators were unable to confirm a bronchodilatory effect of nitrates [226, 227]. Mathys et al [221] even found that airway resistance increased, and that ventilatory functions were impaired by nitroglycerin.

15.3. Nitrates in Acute Respiratory Distress Syndrome

The late stages of acute respiratory distress syndrome (ARDS) are associated with hemodynamic changes, including pulmonary hypertension, increased pulmonary vascular resistance and, in some cases, right ventricular failure. Cardiac output may be decreased, increased or normal. Draxler et al [228] studied the effects of ISDN, given intravenously at infusion rates of 3 – 12 mg/h, on circulation and gas exchange in seven patients with ARDS. All patients were artificially ventilated at the time of the study. Systemic vascular resistance, systemic arterial pressure, heart rate, cardiac output and stroke index were not altered. Mean pulmonary artery pressure was significantly decreased at infusion rates of 6, 9 and 12 mg/h, to 61.7%, 61% and 69.3%, respectively, of the initial value. Pulmonary capillary wedge pressure decreased to 49.6%, 45.5% and 41.5% of the initial value, respectively. Significant decreases of right ventricular stroke work index to about 60% of the initial value were observed at infusion rates of 6 and 9 mg/h.

Thus, administration of ISDN in ARDS caused an improvement in right ventricular function as a result of reductions in preload, afterload and pulmonary blood volume. Slight alterations in the ventilation/perfusion ratio which were also produced by the drug were well tolerated.

This study does not, however, answer questions associated with the long-term effect of ISDN on pulmonary circulation and function.

15.4. Effect of Nitrates on Abdominal Vessels and on the Aorta

Nitrates dilate abdominal vessels. Schneider et al [229] studied 15 patients with coronary artery disease who received oral ISDN, 20 mg six times daily, and placebo for 4 weeks each, in a randomized, double-blind trial. On days 1, 14 and 28 of each period the diameters of the superior mesenteric artery, mesenteric vein, hepatic artery and portal vein were determined by ultrasound. A sublingual tablet of 0.8 mg nitroglycerin was then given to evaluate the dilatory reserve. The absolute diameters did not differ significantly during the two phases. On the 1st day of treatment sublingual nitroglycerin produced a greater increase in diameters in the placebo group compared with minimal increase in the ISDN group. On day 14 the difference between the response to nitroglycerin in both groups was smaller and nonsignificant. This was the case also on day 28. It was concluded that ISDN-induced vasodilation in abdominal vessels is attenuated during sustained therapy. Response of these vessels to nitroglycerin is restored despite long-term ISDN therapy.

Nitroglycerin was shown to relax isolated rings of aorta in vitro. Studies in human subjects have not been performed.

15.5. Nitrates in Peripheral Vascular Diseases

Being mainly venodilators, nitrates might be of only limited benefit in peripheral arterial disease. Nevertheless, attempts to treat peripheral vascular diseases with nitrates have been made over the past 40 years. As early as 1948 it was reported that transcutaneous administration of nitroglycerin, in the form of ointment, produced symptomatic relief in such patients. The diseases treated with nitroglycerin ointment are intermittent claudication, Raynaud's disease, arteriosclerotic gangrene and others [230].

Francis et al [231] studied the effect of nitroglycerin ointment on plethysmographic measurements in the lower extremities. When applied to the right foot the drug produced an increase in pulse volume at least twice that of the left foot in 70% of the patients. When applied to the left foot the drug produced a 2-fold increase in pulse volume in 30% of the patients. These findings indicate a local effect on the circulation as well as a systemic effect.

Testa and coworkers [231a] studied ISDN ointment, 100 mg three times daily, applied directly to the areas where ischemic pain was experienced, in 30 male patients with stable documented intermittent claudication. The symptom-free distance walked (DWA) and the maximum distance reached (MDR) basally and after 1, 3, 6 and 12 months were evaluated by means of treadmill stress tests. After the basal test patients were randomly divided into two groups: placebo group and therapy group (double-blind), and a further treadmill test was performed 1 month later: DWA was 92 ± 24 m versus 297 ± 85 m and MDR 232 ± 53 m versus 506 ± 86 m in the therapy group ($p < 0.01$) and 74 ± 8 m versus 96 ± 15 m and 163 ± 22 m versus 183 ± 23 m in the placebo group, respectively. The investigators then opened the trial and treated all patients, repeating further treadmill tests at 3, 6 and 12 months. The following results were obtained: DWA 84 ± 13 m, 316 ± 63 m, 374 ± 55 m and 452 ± 61 m; MDR 197 ± 29 m, 431 ± 59 m, 514 ± 57 m and 547 ± 49 m, respectively, in basal conditions and after 3, 6 and 12 months of treatment ($p < 0.01$). These results suggest the effectiveness of topical ISDN in improving walking tolerance in patients with intermittent claudication, independently of a placebo effect. Similar results were reported by Biagini et al [232].

At present nitrates are seldom used to treat peripheral vascular disorders.

15.6. Prophylactic Treatment of Phlebitis and Extravasation

Recently a transdermal nitroglycerin patch has been approved for a new indication: prophylactic treatment of phlebitis and extravasation due to venous cannulation for intravenous administration of drugs and fluid.

The main cause of phlebitis and extravasation after intravenous cannulation is probably vasoconstriction, due to mechanical irritation of the vein. Administration of a nitroglycerin patch distal to the affected site can cause vasodilation and continued flow from the cannula, thus avoiding the need for replacement of the infusion site. The therapeutic effect of the patch may be sustained for 3–4 days. Successful treatment can also save money by avoiding the need of reciting the cannula.

It should be remembered that nitroglycerin from the patch is absorbed to the systemic circulation. Headache has been observed in patients receiving this treatment.

16. Non-Cardiovascular Indications for Nitrates

16.1. Achalasia

Achalasia is characterized by loss of peristaltic activity and impairment of relaxation of the lower esophageal sphincter. The sphincter is often hypertensive. This results in esophageal spasm, creating most of the symptoms, signs and complications of achalasia. For about 40 years nitrates have been used to achieve symptomatic relief in this condition. The mechanism of action is relaxation of the esophageal smooth muscle, similarly to the effect of nitrates in vascular walls. Nitrates may be used for this indication by inhalation, as sublingual tablets or as oral spray.

The development of esophageal manometry provided a tool for direct evaluation of the effect of nitrates. Gelfond et al [233] studied the effect of ISDN, 5 mg sublingually, on the lower esophageal sphincter in 24 patients with achalasia. In ten of the patients the sphincter pressure was lowered for 1 hour or more. Nineteen of 23 patients who were treated with ISDN, 5 mg sublingually three times daily (before meals) for 2–19 months, reported a marked relief or disappearance of dysphagia.

Rozen et al [234] also found a lowering of the lower esophageal sphincter pressure by ISDN, 5 mg sublingually, in 15 patients with achalasia. Emptying time of the sphincter, measured by radionuclide methods, was shortened by the drug. These investigators [235] compared the effects of ISDN, 5 mg sublingually, and nifedipine, 20 mg sublingually in patients with achalasia. ISDN was more effective than nifedipine in the acute study. The fall in sphincter pressure was greater and appeared earlier, the esophageal emptying time was shortened to a greater extent, and subjective improvement was greater with ISDN than with nifedipine.

My colleagues and I have recently also shown that ISDN oral spray is effective in lowering the sphincter pressure.

When achalasia is caused by functional obstruction, ISDN appears to be the drug of choice to reduce the sphincter pressure and relieve symptoms.

16.2. Bleeding from Esophageal Varices

Nitrates have been used to prevent recurrent bleeding from esophageal varices. They have been shown to reduce liver blood flow and to lower portal pressure. When combined with beta-blockers, cardiac output is also reduced.

Freeman et al [236] reported that intravenous ISDN produced a significant fall in portal blood pressure in ten patients with esophageal varices. Six of these patients were then treated for 1 month with ISDN, 20 mg four times daily, orally. In each of them the drug significantly lowered portal pressure at the end of this period.

Hallemans et al [237] reported that in ten patients with liver cirrhosis sublingual administration of ISDN, 5 mg, significantly lowered portal pressure. It is yet to be confirmed whether this beneficial effect on portal pressure is presented also as a reduction in frequency of recurrences of bleeding from esophageal varices.

16.3. Glaucoma

As early as 1964 Withworth and Grant [238] reviewed the use of nitrates and nitrites by glaucomatous patients and found no evidence that these agents induced or aggravated an increase in intraocular pressure. However, due to an observation from the early years of the century that nitrates may increase the intraocular pressure in cats they were usually contraindicated in glaucomatous patients. Only recently have their effects been systematically studied in humans. Wizemann et al [239] reportet that ISDN, 80 mg daily, orally, significantly lowered the intraocular pressure in 16 glaucomatous patients. In 1982 Wizemann [240] reported 60 patients with open-angle glaucoma, treated by 80 mg ISDN daily for 2 years, and compared them to patients treated with topical beta-blockers. ISDN had no cardiac or pulmonary side effects and lowered the intraocular pressure to a greater extent than the topical beta-blockers. Patients with systemic arterial hypotension should not receive ISDN.

In patients with closed-angle glaucoma nitrates may be less effective, but there is no formal contraindication to their use.

17. Nitrates in Children and the Elderly

17.1. Nitrates in Infants and Children

Vasodilators have been used extensively in infants and children, mainly to treat acute hemodynamic deterioration in hospitalized patients. Nitroprusside has traditionally been the most widely used vasodilator in the pediatric age group. The main indication for nitroprusside in infants and children is hypertensive emergencies. The drug has the advantage of being a balanced vasodilator with quick onset of action. Recently, however, nitroglycerin and other nitrates have partially replaced nitroprusside. This is because of three factors which limit the use of nitroprusside:

1. The drug is not effective orally.
2. It has a potent hypotensive effect and consequently requires careful monitoring.
3. Prolonged administration may result in thiocyanate toxicity.

Intravenously administered nitroglycerin has recently been suggested as the vasodilator of choice in the postoperative management of pediatric cardiac patients with pulmonary hypertension. Ilbawi et al [241] studied 20 infants and children, aged 4 months to 6 years, who received intravenous nitroglycerin following intracardiac repair of congenital heart diseases. Eleven of the patients had ventricular septal defect, six had atrial septal defect, and three had atrioventricular canal. All of them had a ratio of pulmonary/systemic flow greater than 2.5:1. In 14 patients the preoperative pulmonary arterial pressure was equal to the aortic pressure, and in five patients pulmonary vascular resistance was higher than 6 U. In the 14 patients with elevated preoperative pulmonary arterial pressure intravenous nitroglycerin increased cardiac index from 3.06 ± 0.48 to 3.74 ± 0.45 L min^{-1} m^{-2} and reduced systemic vascular resistance from 1635 ± 217 to 1294 ± 249 dyn sec cm^{-5} and pulmonary vascular resistance from 305 ± 78 to 214 ± 76 dyn sec cm^{-5}. In the six patients with normal preoperative pulmonary artery pressure nitroglycerin increased cardiac index from 3.97 ± 0.61 to 4.30 ± 0.77 L min^{-1} m^{-2} and reduced systemic vascular resistance from 1602 ± 359 to 1137 ± 305 dyn sec cm^{-5}. The changes were more significant in the patients with elevated preoperative pulmonary artery pressure. The investigators concluded that intravenous nitroglycerin is the vasodilator of choice in the postoperative management of the pediatric cardiac patient with pulmonary hypertension. In infants and children nitroglycerin has been given until now only intravenously. The conventional dose is 5 μg/kg per minute. In the future nitrates will probably be given also orally, as spray, or transdermally.

The pediatric experience with ISDN is limited. My colleagues are now performing a dose-finding study of intravenous ISDN in infants and children with heart failure due to endomyocardial diseases. We have also observed clinical improvement in a child with pulmonary edema treated by ISDN oral spray.

17.2. Nitrates in the Elderly

Nitrates are used in elderly patients in all the indications for which they are used in younger ones. The antianginal effect of nitrates in the elderly is comparable to that observed in younger patients. The tolerability of nitrates in the elderly may be superior to that of other classes of antianginal agents, such as beta-blockers and calcium antagonists, although this aspect has not been specifically studied. Comparable efficacy in elderly and young patients was also observed in unstable angina pectoris, acute myocardial ischemia and infarction, congestive heart failure and hypertensive emergencies.

The most important progress in nitrate therapy for the elderly has been the development of slow-release formulations, such as sustained-release oral ISDN or nitroglycerin, IS-5-MN long capsules and transdermal nitroglycerin discs. These formulations have markedly improved patient compliance and allowed use of nitrates for prevention rather than for treatment of angina.

It has been suggested that aging might structurally alter the veins or the venodilatory effects of pharmacologic agents. If so, the optimal doses of nitrates might be different in elderly than in young patients. Gascho et al [241b] studied venous distensibility during control conditions and after the administration of 0.8 mg nitroglycerin spray, in 50 healthy subjects ranging in age from 21 to 78 years. They found that both baseline venous distensibility and the venodilatory response to nitroglycerin decreased with age. These findings suggest that the dose requirement of nitroglycerin might be higher in the elderly.

17.3. Elderly Patients with Congestive Heart Failure

Recently my colleagues and I have demonstrated a difference in the response to nitroglycerin between elderly and young patients with congestive heart failure. In this trial Marchionni and coworkers [241a] studied the effect of age on the vasodilatory properties of intravenous nitroglycerin in patients with congestive heart failure complicating acute myocardial infarction. In all 24 patients, 8 of them younger and 16 older than 65 years of age, the dose of nitroglycerin was titrated to achieve a maximal increase in cardiac output. All effects of nitroglycerin were similar in both age groups except for percentage reduction in pulmonary capillary wedge pressure, which was greater in elderly than in young patients (48 ± 5% versus 27 ± 6%). Thus, the arterial dilatory properties of nitroglycerin are not age-related. This is in accordance with the findings of an experimental study reported by Fleisch and Hooker [242]. However, the venodilatory effect of nitroglycerin was significantly greater in elderly patients than in younger ones. This may result from an age-related decrease in the circulating blood volume.

Nitrates were found effective in patients older than 75 years of age who were included in several studies. For example, Figueras et al [243] evaluated the effect of oral and chewable ISDN in patients with refractory heart failure up to 82 years of age (mean: 67 years). Both formulations reduced pulmonary arterial end-diastolic pressure by more than 10% in 15 patients (responders) and by 10% or less in three others (nonresponders). In the responders both agents produced significant decreases in pulmonary arterial end-diastolic, mean right atrial and mean arterial pressures. Oral ISDN also increased cardiac index and stroke volume.

I have studied five patients aged 73–79 years with severe left heart failure complicating acute myocardial infarction. All had clinical and roentgenographic signs of congestive heart failure and left ventricular filling pressure of 26 mmHg or more at rest. Intravenously administered ISDN reduced the left ventricular filling pressure from 31 ± 2 to 20 ± 2 mmHg and systemic vascular resistance from 2034 ± 181 to 1422 ± 204 dyn sec cm^{-5} and increased cardiac output from 3.0 ± 0.2 to 3.9 ± 0.1 L min^{-1} m^2. The hemodynamic improvement was associated with symptomatic relief.

The demonstration that elderly patients with heart failure respond favorably and, in general, similarly to young patients to nitrates has an important practical implication. In certain countries elderly patients with heart failure are not admitted to coronary care units where treatment with vasodilators is given. The studies quoted in this chapter clearly indicate that such patients can benefit from nitrate therapy.

A common clinical problem in geriatric therapy is the immediate management of elderly patients who come to the clinic or emergency room suffering from tachypnea and dyspnea due to transient exacerbation of congestive heart failure. Having had many previous hospitalizations, these patients often do not have easily accessible veins, and starting an intravenous line may require at least several minutes. Distressed patients also find it difficult to hold a sublingual tablet.

ISDN oral spray can solve this problem as it acts within 1 min and requires no intravenous equipment and only minimal cooperation of the patient. Marmor and coworker (personal communication) studied the effect of ISDN oral spray on respiration in 15 patients, aged 60–84 years, with tachypnea due to acute exacerbation of congestive heart failure. Within 10 min from administration and without additional therapy, one squirt of the spray reduced respiration rate from 41.1 ± 7.6 to 23.5 ± 3.0 per minute. Other clinical parameters of congestive heart failure had also improved. Heart rate and blood pressure were not significantly altered.

ISDN oral spray can improve emergency treatment of congestive heart failure in the elderly.

17.4. Angina Pectoris

Most cardiologists and geriatricians know from their clinical experience that there is no age limit for the antianginal efficacy of nitrates, either for relief of pain or for prophylaxis. However, no specific studies have been performed in the elderly, although elderly patients have been included in many clinical trials. Moreover, very few elderly patients with angina pectoris have been evaluated by exercise test to demonstrate antianginal efficacy of nitrates (or any other antianginal agent). My colleagues and I have initiated such a study, the results of which will be available in 1988.

Nitrates are effective for treatment of unstable angina in the elderly. Lin and Flaherty [244] evaluated the effect of intravenous nitroglycerin, titrated to a fixed hemodynamic end-point, as initial therapy for elderly patients with severe unstable angina pectoris. Five patients, aged 66–76 years, were studied. The drug, at an infusion rate of 12–250 μg/min, completely abolished pain in all five patients.

Curfman et al [84] studied the effect of intravenous nitroglycerin in 14 patients of mean age 65 years, with unstable angina pectoris. The drug reduced the number of spontaneous ischemic episodes from 3.3 ± 0.8 to 1.0 ± 0.3 per 24 h. This effect was similar to that produced by oral ISDN combined with nitroglycerin ointment. As expected in this condition, 40% of patients required coronary artery bypass surgery to completely control the ischemic episodes.

It may be concluded that nitrates are effective in all forms of angina pectoris in the elderly. Further controlled studies are required in elderly patients with chronic stable angina pectoris.

17.5. Transdermal Nitrates

Transdermal nitroglycerin patches are theoretically optimal for elderly patients because they simplify the therapeutic regimen. However, very few elderly patients have been included in studies of transdermal nitroglycerin patches, and in no case have they been analysed separately. No pharmacokinetic studies of transdermal nitroglycerin in the elderly have been reported.

17.6. ISDN Spray in Elderly Patients with Angina Pectoris

Dryness of mouth is more common in elderly patients than in younger ones. This may delay dissolution of sublingual tablets and therefore delay the onset of antianginal effect. ISDN oral spray overcomes this problem as no dissolution is required. My colleagues and I have studied the rapidity of disappearance of pain after ISDN oral spray and compared it with sublingual tablets of ISDN in nine elderly patients with exercise-induced angina pectoris. The patients underwent bicycle ergometry which resulted in typical chest pain associated with ST-segment depression. Immediately upon termination of exercise they received either oral spray or a sublingual tablet. At least 6 h later the study was repeated with the other drug. ISDN in both formulation relieved pain in all patients. The oral spray, 1.25 mg, relieved pain after 61.6 ± 24.4 sec and the tablet, 5 mg, after 112.4 ± 70 sec. The difference was highly significant ($p < 0.0005$).

It was concluded that in elderly patients ISDN spray relieves pain of angina pectoris about twice as rapidly as ISDN sublingual tablet.

C. Nitrate Formulations and Routes of Administration

18. Sublingual Tablets and Oral Spray

For about 100 years the primary indication for nitrate therapy has been the abolition of pain and ischemia during acute episodes of angina pectoris. Sublingual tablets of nitrates have traditionally been used for this indication. Recently several formulations of nitrate spray, which are superior to sublingual tablets in rapidity of onset of effect as well as in several other aspects, have been developed.

18.1. Sublingual Tablets

The oral route is too slow for administration of drugs aimed at relieving acute episodes of angina pectoris, and the intravenous route is inadequate for this indication. Administration of nitrates in the form of sublingual tablets achieves both a rapid appearance of therapeutic plasma levels and avoidance of first-pass hepatic metabolism. As the purpose of this mode of therapy was to relieve pain and not to prevent recurrent anginal episodes, nitroglycerin was used. The pharmacokinetic advantages of other nitrates were considered irrelevant for the indication of acute relief of pain and ischemia. Shortly after achieving its goal the drug was cleared from the plasma. As no reservoir was present, each sublingual tablet of nitroglycerin resulted in a single peak of therapeutic plasma level, and, therefore, the question of tolerance was never raised. Later it was realized that the very short duration of action of nitroglycerin may not be sufficient to protect the patient during periods of effort lasting 15 min or more. Sublingual administration of ISDN was found both to relieve pain as rapidly as sublingual nitroglycerin and provide antianginal protection for 30–90 min. In the study of Willis [245] the duration of action of sublingual ISDN was 6 times longer than that of sublingual nitroglycerin. In present practice both nitroglycerin and ISDN are used for sublingual administration.

An important finding in nitrate research was that sublingual tablets produce a therapeutic effect even when given to patients who are already being treated with long-acting nitrates. Despite the finding of some cross-tolerance sublingual administration of nitroglycerin relieves angina and ischemia in patients chronically treated with oral ISDN, and sublingual ISDN relieves angina even in patients chronically treated with transdermal nitroglycerin.

The use of sublingual nitrate tablets has stood the test of time in worldwide clinical practice. Their effects on ischemia, determinants of myocardial oxygen demand and coronary arteries have been evaluated in numerous clinical studies. Rudolph et al [246] studied the influence of ISDN, 5 mg sublingually, on the regional myocardial blood flow and ventricular wall movement. They found an increase of 20% in perfusion in poststenotic underperfused regions and a decrease of 14% in normal myocardial tissue. This was associated with enhanced contractility of hypokinetic left ventricular myocardial segments. Brown et al [247] studied 120 stenosed segments in 64 patients. Both ISDN and nitroglycerin dilated high-grade stenoses by 38% and medium-grade stenoses by 25%. These results are comparable to those achieved with intravenous administration of nitrates.

The hemodynamic effects of sublingual nitrates in angina pectoris were studied by several investigators, in the early 1970s. For example, Sweatman et al [248] assessed the hemodynamic effects of sublingual ISDN and nitroglycerin in 27 patients with angina pectoris. ISDN produced a significant decrease in pulmonary capillary wedge pressure, cardiac index, stroke index, stroke work index and mean arterial pressure measured 1 h after administration of the drug. Nitroglycerin did not produce any sustained significant changes. Baxter et al [249] studied the hemodynamic effects of ISDN, 5 mg sublingually, in patients with acute myocardial infarction and elevated left ventricular filling pressure. A 37% reduction in filling pressure within 10 min following drug administration was observed. This effect persisted for 1–2 h and in some patients even longer. Therefore sublingual ISDN is suitable for patients with acute myocardial infarction and high left ventricular filling pressure and may be considered an alternative therapy for those who do not respond to diuretics. (Kiely et al [250] reported that a reduction in left ventricular filling pressure cannot be obtained with furosemide in 50% of patients with acute myocardial infarction.)

Sublingual nitrates have become the standard therapy for acute episodes of angina pectoris. Their use is considered a marker for the presence of this condition. In most long-term studies of the efficacy of antianginal agents only sublingual nitroglycerin is allowed in addition to the drug tested, and the reduction in nitroglycerin consumption is considered an indication of the antianginal efficacy of the tested drug.

The effect of sublingual nitrates is exerted within 2–5 min. For many years it has been the most rapid way to administer nitrates. Only recently it has been shown that ISDN and nitroglycerin sprays exert their effect more rapidly than the corresponding sublingual tablets.

Mononitrates are not used sublingually because their special pharmacokinetic properties have no advantage for rapid administration.

18.2. Nitrate Oral Spray

Nitrates are widely used for the treatment (in the form of sublingual tablets) and prevention (in the form of oral tablets and transdermal delivery systems) of attacks of angina pectoris. Even the traditionally most rapid method of administration, the sublingual tablet, takes at least 2 min until the onset of effect. Usually this takes some more time because 2 min are the time required from the moment of dissolution by nitrates, administered in the form of sublingual tablets, to reach peak plasma levels [251]. An additional period is, however, required for dissolution of the tablet which must take place, at least partially, before absorption begins. This period shows interpatient variability and further delays the onset of action.

It is a common practice to administer nitrates in the form of sublingual tablets before effort which is known to cause chest pain in patients with angina pectoris. Due to the reasons discussed, the onset of this effect takes several minutes. Although this is usually sufficient for most patients with angina pectoris, a more rapid method of administration is desirable.

Several forms of oral spray, in which the active compound (ISDN or nitroglycerin) is kept as a solution in a spray container, are available. These were planned to avoid the delay required for dissolution of sublingual tablets. ISDN spray was found effective in a variety of cardiovascular diseases: it relieves pain in patients with chronic stable or unstable angina pectoris and produces hemodynamic and symptomatic improvement in patients with acute exacerbation of congestive heart failure and pulmonary edema.

18.2.1. Nitrate Oral Spray in Angina Pectoris

The most important field in which nitrate oral spray may be beneficial is in the treatment of acute episodes of angina pectoris, where the rapidity of onset of effect is essential. Several studies have revealed that ISDN oral spray exerts an antianginal and anti-ischemic effect of very rapid onset in angina pectoris.

My colleagues and I assessed the antianginal effect of ISDN spray in patients with exercise-induced angina pectoris in order to determine and quantify its time course. Ten patients (mean age 58.9 ± 6.2 years) were included in the study. All had chronic stable angina pectoris. The length of time to onset of angina and to appearance of a 1.0-mm ST-segment depression was at least 0.08 sec; the time to disappearance of pain and to ST-segment depression after discontinuation of exercise was measured. ISDN spray delayed the onset of angina and electrocardiographic ischemic changes in all patients. Exercise time to pain was 5.1 ± 1.4 min with placebo and 7.2 ± 1.3 min with ISDN ($p < 0.001$). Time to appearance of ST-segment depression was 7.1 ± 1.5 min with placebo and 10.2 ± 1.2 min with ISDN. Time to disappearance of pain after discontinuation of exercise was shortened from 3.2 ± 0.8 min on placebo to 2.1 ± 0.8 min on ISDN ($p < 0.001$). Time to disappearance of ST-segment depression was shortened from 4.2 ± 0.6 to 2.5 ± 0.8 min, respectively, ($p < 0.005$). Two patients experienced transient headache and one patient had transient dizziness with slight decrease in blood pressure after ISDN. All patients completed the study without experiencing major adverse effects.

Our previous experience with ISDN spray suggests that this may be superior to sublingual tablets of nitrates as it acts more rapidly. Previously we had observed that ISDN spray exerts a beneficial hemodynamic effect within 30 sec in patients with acute pulmonary edema, and that the spray relieves exercise-induced chest pain and electrocardiographic changes 30%–50% more rapidly than sublingual tablets of ISDN. It may be of interest to point out that the onset of effect is more rapid with ISDN spray than with intravenous infusion. This is because of the time required to start an intravenous line [252, 253].

Therefore, this formulation has been chosen for evaluation as pre-effort therapy in patients with exercise-induced angina pectoris. Due to the rapid onset of action the spray is especially suitable for use immediately before initiation of exercise, and the patient does not have to wait until dissolution of the tablet.

One study, described above, revealed that ISDN spray, administered immediately before the initiation of exercise, increased exercise tolerance in patients with chronic stable angina pectoris. The drug increased exercise duration by about 40%. This is comparable to the results achieved with sublingual nitroglycerin tablets [254–257] and nitroglycerin oral spray [258]. This increase was associated with the usual cardiocirculatory effects of nitrates.

It is important to note that ISDN did not significantly lower diastolic blood pressure. When considerable autoregulatory coronary dilation has been achieved, as might be expected in patients with coronary artery disease who exercise to myocardial ischemia, myocardial flow distal to a fixed coronary stenosis has been shown to be directly related to perfusion pressure [259, 260].

In the presence of peripheral vasodilation accompanying exercise a vasodilator which excessively lowers diastolic blood pressure may impair coronary perfusion. As ISDN lowers the left ventricular filling pressure, myocardial flow may increase in relation to a given perfusion pressure.

ISDN also significantly shortens the duration of chest pain and ST-segment depression. In this respect the effect of ISDN spray given as prophylaxis immediately before exercise includes the effects that are achieved by nitrates given as treatment at the onset of pain.

Kimchi et al [258] have shown that nitroglycerin oral spray is an effective prophylactic for exercise-induced angina pectoris. The drug delayed or abolished the appearance of chest pain and ST-segment depression. Patients achieved a higher heart rate at peak exercise with nitroglycerin spray, associated with a decline in the maximal exercise-induced ST-segment depression. Our results with ISDN are in accordance with these reports on the effects of nitroglycerin. However, ISDN has a longer half-life than nitroglycerin [261, 262]. Hence it should be theoretically superior in its duration of action. Therefore ISDN spray, which combines very rapid onset of action with relatively prolonged efficacy, has advantages as pre-exercise prophylactic in patients with chronic stable angina pectoris.

Another difference between the studies is that Kimchi et al [258] gave the drug 2 min prior to initiation of exercise. This early administration is required for sublingual tablets but, in our opinion, is not necessary for the spray. The rapidity of action of the spray allows administration immediately at the initiation of exercise, as we carried out in our study.

The spray has a further theoretical advantage over tablets, in that the active drug is protected in the container while tablets are repeatedly exposed when the bottle is opened. This may result in chemical degradation of the tablets. In summary, ISDN spray is an effective prophylactic agent for pre-exercise use in patients with chronic stable angina pectoris. It not only increases the duration of exercise and delays the onset of pain and ischemic electrocardiographic changes but also shortens the duration of these presentations of ischemia when they do develop.

To directly demonstrate the superiority of spray it is important to perform comparative studies with sublingual tablets of nitroglycerin or ISDN. My colleagues and I [263] have performed such a comparative study. We compared the efficacy and rapidity of relief of pain by ISDN spray and ISDN sublingual tablets in ten patients with exercise-induced chronic stable angina pectoris. The evaluation was by a randomized, crossover trial involving bicycle exercise testing. In all patients exercise was terminated because of chest pain resembling that which they usually suffered, associated with electrocardiographic ST-segment depression. Patients received the drugs immediately after termination of exercise. Studies with the two therapeutic forms were performed at different days or at least 6 h apart. They received no other cardioactive drugs in the week prior to the study. ISDN oral spray relieved pain within 2.1 ± 0.9 min from administration, and ST-segment depression within 2.05 ± 0.8 min. The corresponding periods for ISDN sublingual tablets were 3.04 ± 0.68 and 5.48 ± 0.7 min, respectively. The difference was statistically significant for both parameters ($p < 0.005$).

In summary, ISDN oral spray relieved angina and ischemia more rapidly than ISDN tablets. ISDN spray might, therefore, be preferred for immediate relief of pain in patients with angina pectoris. These findings are in accordance with those of a pharmacokinetic study which showed that therapeutic plasma levels of ISDN are achieved several minutes earlier with spray than with a sublingual tablet [264].

Grüntzig [265] studied the effect of ISDN spray on a major parameter affected by nitrates – the left ventricular end-diastolic pressure. The study was performed during coronary catheterization in 26 patients with coronary artery disease. Left ventricular pressure was measured during atrial pacing at increasing rates. When chest pain developed, the patients received either a sublingual placebo capsule or placebo oral spray, and 15 sec later the pacing rate was lowered to the initial value. This sequence was repeated then with ISDN spray, 1.25 mg, and 0.8 mg sublingual nitroglycerin (13 patients in each group). Atrial pacing up to pain and then slowing down resulted in an expected significant rise in left ventricular end-diastolic pressure when placebo was given. Nitroglycerin had no effect on this pressure within the first 15 sec. In contrast, ISDN oral spray prevented the rise in left ventricular end-diastolic pressure. These findings confirm that ISDN oral spray acts more rapidly than sublingual nitroglycerin.

18.2.2. Nitrate Oral Spray in Congestive Heart Failure

Naturally, the superiority of nitrate oral spray is an immediate treatment of acute episodes of the diseases for which nitrates are used. In congestive heart failure an acute episode is usually presented as acute pulmonary edema. Nitrates have substantially improved the management of acute pulmonary edema. The nitrate-induced hemodynamic improvement in this condition results mainly from venodilation and is accompanied by clinical improvement [251–256].

As the rapidity of drug administration plays a major role in the management of acute pulmonary edema, the usual routes of administration of nitrates are the intravenous and sublingual ones. However, intravenous administration may require a few minutes. Sublingual administration of a tablet of nitrate is often not practical in very distressed patients. Moreover, dissolution of the tablets takes about 2 min, before significant absorption begins [257].

ISDN spray appears to be the optimal method for very rapid treatment. My colleagues and I studied two patients with acute pulmonary edema in whom hemodynamic improvement started within seconds after administration of ISDN spray. The patients, both having chronic congestive heart failure, were studied by Swan-Ganz catheter for hemodynamic evaluation. Both developed acute pulmonary edema while the catheter was in place and during chronic treatment with diuretics only. In both ISDN spray produced immediate hemodynamic improvement, evident by decreases in pulmonary artery, capillary wedge and right atrial pressures and increases in cardiac output, within 60 sec. Further hemodynamic improvement, associated with symptomatic improvement, was observed at 5 min. In one of the patients recurrent episodes of pulmonary edema similarly responded to ISDN spray, but a response to sublingual ISDN tablet was observed only at 10 min.

It can be concluded that ISDN spray is important for the very early treatment of acute pulmonary edema and may increase the rapidity of response. It may also be important in cases of very distressed patients who can not take sublingual nitrates.

In a larger and more recent study my colleagues and I have studied the hemodynamic effects of ISDN spray in eight patients with acute deterioration of chronic congestive heart failure. All cardiovascular drugs were discontinued at least 3 days prior to the study and a Swan-Ganz catheter was introduced. During hospitalization the patients developed symptoms and hemodynamic signs of acute deterioration of the chronic condition. All were tachypneic, dyspneic and orthopneic. One squirt of ISDN spray (equal to 1.25 mg of ISDN) produced hemodynamic improvement which became evident in all patients within 1 min and reached a peak within 5.2 min from administration. The drug decreased the pulmonary arterial pressure from $57 \pm 13/29 \pm 10$ mmHg to $40 \pm 7/22 \pm 5$ mmHg and pulmonary capillary wedge pressure from 25 ± 10 mmHg to 17 ± 6 mmHg and increased cardiac index from 2.1 to 3.0 L min^{-1} m^{-2}. Systemic and pulmonary vascular resistances were reduced. All the changes were statistically significant. They were associated with symptomatic relief. That the hemodynamic effect was achieved more rapidly than the starting of an intravenous line, indicates that ISDN spray may be beneficial for emergency treatment of acute deterioration in chronic congestive heart failure.

A new formulation of aqueous solution of isosorbide dinitrate (ISDN) spray is now being developed. We compared the magnitude and time course of the hemodynamic effect of the new formulation to those of sublingual nitroglycerin (NTG) tablets in 12 patients with chronic congestive heart failure. The patients received, in a random order, ISDN spray, 2.5 mg, or sublingual NTG 0.8 mg. Hemodynamic measurements were performed before and at 1, 3, 5, 10, 20, 30 and 60 min after each drug. The second drug was given only after return of the hemodynamic parameters to baseline, plus a washout period of 2 h. The hemodynamic variables measured were comparable at baseline. Both drugs produced hemodynamic improvement including a decrease in pulmonary capillary wedge pressure, (PCWP), right atrial pressure (RAP) and systemic and pulmonary vascular resistances (Table 1). Only ISDN spray significantly increased cardiac output. The onset of action of ISDN spray was significantly more rapid than that of NTG. This was primarily evident in the decrease in PCWP. With ISDN spray the decrease started at 1 min after administration and at 3 min a decrease of 8.6 mmHg was already found. The corresponding value for sublingual NTG was 1.6 mmHg. The difference was highly significant ($p < 0.02$). The peak effect of ISDN spray on PCWP and RAP was greater than that of NTG.

In conclusion: the onset of the hemodynamic effect of the new formulation of ISDN spray is much more rapid than that of sublingual nitroglycerin tablets. At the doses used, the magnitude of the effect of the ISDN spray on some of the hemodynamic variables is greater than that of sublingual NTG.

We have recently performed a dose-finding study of the hemodynamic effect of the new spray in patients with chronic congestive heart failure. Twelve patients with chronic congestive heart failure were studied. Doses of 1.25, 2.5, 5.0 mg and placebo, as one squirt, were randomly given to all patients. Hemodynamic measurements were performed by a Swan-Ganz catheter before and at 30 sec, 1, 5, 10, 20 and 30 min after drug administration and every 30 min thereafter, until return of hemodynamic variables to baseline. Hemodynamic improvement evident as decreases in right-sided pressures and an increase in cardiac output was observed within 1 minute from administration of ISDN spray, and peaked at 5 min. Near maximal effect was achieved by the 2.5 mg dose. In conclusion, 2.5 mg of ISDN spray (new formulation) produces rapid, near maximal hemodynamic improvement in patients with congestive heart failure.

Toman et al [257a] studied the hemodynamic effects of ISDN spray, 2.5 mg, and sublingual nitroglycerin, 0.5 mg, in 10 patients with acute myocardial infarction. Both drugs reduced the right atrial and pulmonary arterial pressures and the right ventricular stroke work index. The maximal effect was observed 10.4 min after ISDN spray and 10.3 min after nitroglycerin. The effect disappeared 30 min after nitroglycerin and 60 min after ISDN. As the spray application may be less stressful than a sublingual tablet, the ISDN spray is a useful extension of the therapeutic armamentarium in patients with acute myocardial infarction.

Table 1. Hemodynamic variables before and at 1, 3, 5, 10, 20, 30 and 60 minutes after drug administration (values expressed as mean ± SD)

Time	*HR*	*MAP*	*PCWP*	*RA*	*CO*	*SVR*
Isosorbide dinitrate spray						
0′	83.5 ± 11.3	94.2 ± 5.8	25.7 ± 5.4	13.6 ± 3.2	3.7 ± 0.3	1741 ± 170
1′	84.7 ± 6.0	93.5 ± 8.9	24.0 ± 4.6	13.3 ± 3.4	3.8 ± 0.3	1694 ± 210
3′	92.2* ± 2.8	86.7* ± 6.0	17.1*,** ± 2.5	9.1* ± 2.3*	3.9 ± 0.3	1606*,* ± 160
5′	93.1* ± 3.0	84.3* ± 5.2	15.4*,** ± 2.1	7.7* ± 2.6	4.0 ± 0.3	1520*,* ± 157
10′	91.2* ± 3.1	85.8* ± 4.6	13.3*,** ± 2.3	6.3*,** ± 1.5	4.1* ± 0.4	1570* ± 177
20′	90.7* ± 2.4	87.8* ± 5.4	16.8* ± 2.0	9.6* ± 3.1	4.1* ± 0.3	1535* ± 199
30′	88.2 ± 3.0	91.1 ± 6.8	17.8* ± 2.0	9.2* ± 2.6	3.9 ± 0.3	1681 ± 214
60′	88.2 ± 5.2	90.6 ± 6.8	18.0* ± 1.7	9.9 ± 2.1	4.01 ± 0.3	1638 ± 203
Sublingual nitroglycerin						
0′	82.4 ± 11.9	95.3 ± 6.4	25.4 ± 4.9	14.2 ± 3.2	3.7 ± 0.3	1765 ± 200
1′	82.8 ± 7.2	92.6 ± 5.8	25.0 ± 4.8	13.8 ± 3.3	3.7 ± 0.3	1703 ± 144
3′	87.2 ± 6.3	92.5 ± 7.6	23.8 ± 6.0	11.9 ± 3.2	3.7 ± 0.3	1768 ± 206
5′	88.5* ± 4.2	89.4* ± 7.8	20.7* ± 6.6	10.7* ± 3.7	3.8 ± 0.3	1681 ± 188
10′	93.1* ± 3.8	86.2 ± 4.4	18.9* ± 4.5	9.7 ± 3.7	3.9 ± 0.3	1568* ± 175
20′	92.0* ± 3.5	87.6 ± 5.5	18.3* ± 4.5	9.5* ± 3.7	3.9 ± 0.3	1608* ± 169
30′	90.3* ± 2.8	89.4 ± 6.2	18.9 ± 3.5	9.7 ± 2.7	3.8 ± 0.2	1674 ± 164
60′	88.0* ± 4.8	91.7 ± 6.7	22.1* ± 3.4	12.4 ± 3.2	3.8 ± 0.4	1695 ± 213

HR: Heart rate (beats/min); *MAP:* Mean systemic arterial pressure (mmHg); *PCWP:* Pulmonary capillary wedge pressure (mmHg); *RA:* Right atrial pressure (mmHg); *CO:* Cardiac output (L/min); *SVR:* Systemic vascular resistance (dynes · sec · cm^{-1}); * Significance of difference from baseline ($p < 0.05$); ** Significance of difference between ISDN spray and NTG ($p < 0.05$)

19. Intravenous Administration of Nitrates

Intravenous administration of nitrates is the most important application form of these agents in emergency situations. The introduction of formulations of nitroglycerin, ISDN and IS-5-MN for intravenous administration has changed the face of acute coronary care. The history of intensive coronary care units can be divided into three periods:

1. The first decade, when arrhythmias were effectively detected and treated, but little could be offered to treat hemodynamic deterioration.
2. The second decade, when intravenous nitrates, together with the modern sympathomimetic amines, provided a therapeutic tool for acute hemodynamic deterioration.
3. The third period, which has just begun, is the era of interventional cardiology. It has created new indications for intravenous nitrates as adjuncts to percutaneous transluminal coronary angioplasty and thrombolysis.

Nitroprusside was the first intravenously used vasodilator, but serious problems have led to its replacement by intravenous nitroglycerin and ISDN. Recently IS-5-MN also has become available for intravenous use.

Intravenous administration of nitrates has several advantages over other forms of administration:

1. The onset of effect is very rapid, within minutes.
2. After steady state is achieved, constant plasma levels can be maintained with a constant infusion rate.
3. The dose may be titrated rapidly and easily.
4. First-pass hepatic metabolism is avoided.
5. The effect disappears rapidly after discontinuation of the infusion.

At present intravenous nitrates are used for treatment of coronary spasm and other forms of unstable angina pectoris, acute-onset congestive heart failure complicating acute myocardial infarction, acute exacerbation of chronic congestive heart failure and complications of cardiac surgery. Some authorities use intravenous nitrates also for reduction of infarct size, although the efficacy in nitrates for this indication is yet to be confirmed. The use of intravenous nitrates for these indications is discussed in detail in the chapters on angina pectoris, acute myocardial infarction and congestive heart failure. The formulations used intravenously may also be injected into the coronary arteries. This is discussed in the following chapter. The use of intravenous IS-5-MN is discussed in the chapter on mononitrates.

20. Intracoronary Administration of Nitrates

Until 1975 nitrates were given directly into the coronary arteries only for clinical research. This was performed to evaluate directly the effects of nitrates on the coronary arteries and myocardial contractility and to differentiate direct from systemic mechanisms of the effects of nitrates. In the past decade, however, intracoronary administration of nitrates, mainly nitroglycerin, has frequently been used as part of diagnostic and therapeutic procedures, including cardiac catheterization, thrombolysis and percutaneous transluminal angioplasty.

20.1. Intracoronary Nitroglycerin in Coronary Spasm

One of the most important indications for intracoronary nitroglycerin is coronary spasm – either spontaneous spasm refractory to conventional methods of nitrate administration or spasm induced during ergonovine test, coronary catheterization or angioplasty. Nitroglycerin probably has a direct relaxant effect on the smooth muscle of the spastic coronary segment.

It is not known why sublingual and intravenous nitroglycerin fail to relieve coronary spasm in some patients. One explanation is that spasm reduces blood flow to the vasa vasorum of the stenotic segment and prevents the drug from reaching the area. Another possibility is sympathetic stimulation, resulting from nitroglycerin-induced systemic vasodilation. Sympathetic stimulation can aggravate coronary spasm. Intracoronary administration avoids the systemic effects and also results in better approach of the drug to the spastic segment.

All cardiologists performing coronary angiography are familiar with the efficacy of intracoronary nitroglycerin in relieving coronary spasm induced by the catheter. It is difficult, however, to perform controlled studies in this condition. Buxton et al [82] reported that intracoronary nitroglycerin relieved ergonovine-induced coronary spasm refractory to other forms of administration of the drug. Pepine et al [83] reported that intracoronary nitroglycerin relieved spontaneously occurring coronary spasm, resistant to sublingual and intravenous administration of nitroglycerin, in both angiographically patent and stenotic coronary arteries.

20.2. Hemodynamic Effects of Intracoronary Nitrates

The doses of nitroglycerin which are usually used for intracoronary administration are too low to produce systemic hemodynamic effects. Feldman et al [266] found no change in heart rate by doses of 5–250 μg despite slight reductions in systemic arterial pressure. Kern et al [267] found no change in heart rate at doses of 50–300 μg, despite reductions of 6% in mean systemic arterial pressure. However, higher doses of intracoronary nitroglycerin produced systemic responses.

Feldman et al [15] reported that 450 μg resulted in an increase of 5 beats/min or more in heart rate and a decrease of more than 10 mmHg in mean arterial pressure.

Hood et al [268] reported that left ventricular pressures and volumes were not significantly altered by intracoronary nitroglycerin, but that the dp/dt was consistently increased. Intracoronary nitroglycerin also did not alter left ventricular stiffness as it did not prevent vascular engorgement in the ventricular wall.

Rutsch and Schmutzler [269] studied the effect of intracoronary administration of ISDN, 50 μg. In patients with mild coronary artery disease the drug increased stroke volume, left ventricular stroke work index, myocardial contractility and myocardial oxygen consumption, suggesting a positive inotropic effect combined with a decrease in coronary vascular resistance and an increase in coronary blood flow. In patients with severe coronary artery disease ISDN produced opposite effects to those observed in mild cases. The investigators suggested that the direct effects of nitrates are of little significance during systemic administration.

20.3. Effects on Coronary Blood Flow

Intracoronary administration of nitrates results in coronary vasodilation and an increase in coronary blood flow. Maximal dilatory effect has been observed with doses of 450 μg [15, 266]. However, a near-maximal effect has already been observed at doses of 200 μg, with only slight additional dilatory effect at higher doses [267]. It is possible that the higher doses of nitroglycerin do have a greater direct dilatory effect, but that it is countered by a systemic effect. The systemic effects of nitrates result in a reduction of myocardial oxygen demand and, via autoregulatory mechanisms, a decrease in coronary blood flow which opposes the direct dilatory effect of nitrates. Another possibility is that the reduction of systemic arterial pressure by nitrates does not allow their direct coronary dilatory effect to be expressed as an increase in coronary blood flow.

Intracoronary nitroglycerin can increase coronary blood flow in human subjects without coronary artery disease. In 1976 Ganz and Marcus [270] reported that intracoronary nitroglycerin, 75 μg, increased coronary blood flow by 84% in five patients. Some years later Winniford et al [271] reported almost similar results with 100 μg intracoronary nitroglycerin in three patients without coronary artery disease.

Smaller increases in coronary blood flow have been observed in patients with coronary artery disease. A recent study was reported by Liu et al [272], who observed a 14% increase in mean coronary blood flow by 40 μg intracoronary nitroglycerin in nine patients. Other investigators reported mean increases of 14%–46%. In one of the older studies a decrease in coronary flow was observed [270]. The most recent study, of 19 patients with coronary artery disease, was reported in 1986 by Kern et al [267], who observed a 53% ± 25% increase in coronary blood flow by 200 μg intracoronary nitroglycerin.

Miller et al [273] studied the effects of intracoronary nitroglycerin on myocardial blood flow to collateral dependent myocardial zones in three groups of patients: group 1, with normal coronary arteries; group 2, with narrowing of more than 70% in the left anterior coronary arteries; and group 3 with total occlusion of the left anterior descending coronary artery, anterior wall dyssynergy and collateral supply. Nitroglycerin was infused intracoronarily at a dose of 200 μg. The drug increased blood flow in the normal coronary arteries. This effect was preserved but significantly attenuated in infarcted zones supplied by collateral vessels. Nevertheless intracoronary nitroglycerin may relieve ischemia during invasive procedures likely to alter mean blood flow to collaterally dependent regions.

20.4. Nitrates in Percutaneous Transluminal Coronary Angioplasty

Intracoronary administration of nitroglycerin is commonly used as part of routine percutaneous transluminal coronary angioplasty. It often precedes and facilitates manipulation of the balloon catheter and may enhance identification of complications during the procedure or immediately after it.

Dilation time is one of the determinants of the success rate of percutaneous transluminal coronary angioplasty. It has been suggested that intracoronary injection of nitroglycerin can prolong dilation time and thus increase ischemic tolerance. Erbel et al [274] studied the effect of intracoronary nitroglycerin on dilation time in ten patients with unstable angina pectoris and single vessel coronary disease. Long-acting antianginal medications were discontinued 12 h before the procedure. Two dilations were performed within 5 min, each time continuing until the development of angina pectoris or ventricular arrhythmias. Nitroglycerin, 0.2 mg, was then injected into the coronary artery, and dilation was repeated after 1.5 and 10 min. The drug increased ischemic tolerance in seven of ten patients, as evidenced by prolongation of dilation time to development of angina pectoris or ventricular arrhythmias.

Some investigators have suggested that the combination of nitroglycerin with calcium antagonists may be more effective than nitroglycerin alone in prolongation of dilation time.

Intracoronary nitroglycerin may be effective in asymptomatic coronary spasm not associated with electrocardiographic changes, as well as in severely symptomatic cases of coronary spasm [267].

Hermann et al [275] compared the effects of 0.2 mg nitroglycerin injected into the coronary arteries to that of the same dose given intravenously in 66 patients undergoing coronary angioplasty. Prior to angioplasty all patients received nifedipine and nitrates sublingually. Both intracoronary and intravenous nitroglycerin improved myocardial ischemia tolerance during angioplasty, and in both cases this was attributed to the systemic effects of the drug. These findings suggest that routine intracoronary administration of nitroglycerin during angioplasty may not be necessary, and that systemic administration is sufficient. Crake et al [276] studied the effect and the mechanism of action of nitrates on myocardial ischemia during coronary angioplasty in 19 patients. They found that the duration of balloon inflation to the onset of ST-segment depression was prolonged by 48% by the intracoronary administration of ISDN, and that the duration to a 1-mm ST-segment depression was prolonged by 28%. Systolic blood pressure fell from 118 $\pm$ 3.2 to 111 $\pm$ 4.1 mmHg after administration of ISDN, but there was no change in intracoronary distal occlusion pressure. It may be concluded that ISDN delays the onset of myocardial ischemia during coronary angioplasty. Recently it was shown that the vasodilatory responses to intracoronary nitroglycerin after transluminal angioplasty are, at least transiently, impaired and do not appear to be adequate as initial quantitative measures of revascularisation [277]. This point, however, deserves further evaluation.

21. Oral Administration of Nitrates

Until 1980 oral administration of nitrates was the only practical way to achieve therapeutic plasma levels of nitrates for long periods in ambulatory patients. For two decades oral ISDN has been the almost exclusive drug for this indication. It allowed long-term prophylaxis of angina pectoris and treatment of congestive heart failure. Although in 1972 it was questioned whether a sufficient amount of ISDN can avoid first-pass hepatic elimination and reach the systemic circulation, it was shown shortly thereafter that this is the case, and ISDN regained its position as first-choice nitrate for chronic oral administration. Recently other nitrates have been used successfully for long-term oral administration. It has been noted that IS-5-MN has some pharmacokinetic advantages for oral administration. Oral sustained-release nitroglycerin formulations have also been used.

Studies on ISDN and sustained-release nitroglycerin are discussed in the chapters on the various indications. Some of the important studies which emphasize the advantages of this form of therapy are discussed below. The effects of orally administered ISDN on angina, ischemia, hemodynamics and coronary arteries have been studied by numerous investigators.

Udhoji and Heng [278] studied the hemodynamic effects of oral sustained-release ISDN (40 or 80 mg) in ten patients with chronic stable angina pectoris. They found a reduction in mean arterial pressure, cardiac index, pulmonary capillary wedge pressure and pulmonary arterial pressure which appeared within 2 h and was sustained for more than 12 h. Similar results were found also in a study of 20 patients with chronic stable angina pectoris reported by Flugelman et al [279]. Additional hemodynamic studies of oral ISDN are discussed in the chapters on angina pectoris and congestive heart failure.

The goal of long-term treatment with oral ISDN is to increase exercise tolerance and reduce the frequency of anginal episodes. This effect of the drug has been demonstrated by both short-term and long-term studies. Danahy et al [11] determined the effect of high-dose oral ISDN on exercise time to angina in 21 patients with documented coronary atherosclerotic heart disease. The patients were tested with 0.4 mg of sublingual nitroglycerin, subsequently with oral ISDN, mean dose 29 mg, and oral placebo in a randomized, double-blind crossover manner. A consistent and prolonged hemodynamic response to ISDN was demonstrated, associated with an increase in exercise time for up to 5 h. These investigators reported also the results of a randomized, double-blind crossover trial of ISDN used over an average period of 5.6 months in patients with chronic stable angina pectoris. Exercise time increased by 58% 1 h after the initial dose of ISDN. A similar increase 1 h after a dose was observed also during chronic oral treatment.

Ohlmeier et al [280] studied 21 patients with coronary artery disease, documented by coronary angiography. The patients received ISDN slow-release, 120 mg once daily, and a symptom-specific exercise test was performed at 2, 12 and 24 h after medication. After 1 week of therapy similar tests were performed. At 2 and 12 h after acute administration working capacity increased to 220% and 139%, respectively. After 24 h there was no statistically significant effect. The ST-segment depression was reduced at 2 and 12 h. After 1 week of therapy there was a slight reduction of action but no development of tolerance. The problem of tolerance can be solved with ISDN itself, and it is not required to replace it by other nitrates. It was shown that long-lasting, unfluctuating, high plasma concentrations (above 300 ng/ml of the active metabolite IS-5-MN) cause marked attenuation of the initial effect. Dosage regimens which provide intervals with low plasma levels were shown to prevent tolerance. This is the case with low-dose regimens such as ISDN 20 mg, two to three times daily. Sustained efficacy may also be obtained with a once-daily 120-mg ISDN sustained-release dose in the morning, which results in plasma levels of 600 ng/ml IS-5-MN throughout the active hours of the drug, with an interval at a plasma level of 100 ng/ml during the night. This regimen was tested for 4 weeks with no loss of efficacy [281].

Some of the centers with widest experience with nitrates appear to be less concerned over the problem of tolerance, even when conventional regimens of low as well as high doses of oral nitrates are used. One of the centers with greatest experience in the field is the cardiology department of the University of Frankfurt/M. Schneider and Kaltenbach from this department have recently reported the results on a study of the treatment of 15 patients with coronary heart disease and exertion-related angina pectoris using ISDN tablets in 4 doses (six times daily: 5 mg, 20 mg, 40 mg, 80 mg). The drugs were given at a randomized sequence, for 1 week each. The treatment resulted in a dose-related improvement of ischemia (exercise-induced ST-segment depression): 30 mg/day, 24%; 120 mg/day, 40%; 240 mg/day, 60%; 480 mg/day, 74%. Continuation of treatment for another 4 weeks with the highest dose resulted in a slight attenuation in antianginal activity, with a 55% improvement of ST-segment depression. The frequency of anginal episodes was also decreased in a dose-related manner [282]. These investigators found a similar dose-related antianginal activity also for oral sustained-release nitroglycerin in 12 patients enrolled in a double-blind crossover trial: 2.6 mg, 23% (nonsignificant); 6.5 mg, 38%; 10 mg, 55%; 20 mg, 74% ($p < 0.0001$).

These findings indicate that even though tolerance to hemodynamic parameters may rapidly develop, the antianginal activity of orally administered nitrates is maintained during sustained therapy.

However, even in patients with impaired left ventricular function, in whom the improvement in hemodynamic parameters should be maintained for sustained efficacy it was shown that oral nitrates can achieve this goal. Wortmann et al [283] studied the initial and chronic effects of ISDN, 120 mg sustained-release, once daily, in nine patients with coronary artery disease and impaired left ventricular function. At intraindividually identical workloads the pulmonary capillary wedge pressure was reduced from 32.5 ± 9.5 to 19.7 ± 9.8 mmHg after the first dose of ISDN. This reduction was fully maintained during chronic therapy. With the first dose of ISDN cardiac index at maximum workload increased from 6.0 ± 1.2 to 6.8 ± 1.3 L $min^{-1}m^{-2}$, while during long-term therapy it increased from 5.3 ± 1.3 to 6.6 ± 1.1 L min^{-1}. Exercise capacity increased from 414 to 686 watts/min after the first dose and further increased to 772 watts/min during long-term therapy. ST-segment depression, measured as the sum of this parameter in all 12 standard ECG leads, decreased from 0.63 to 0.11 mV after the first dose and to 0.16 mV during long-term therapy. The improvement in left ventricular ejection fraction and in the response of the coronary arteries to ergonovine maleate were also maintained during long-term therapy [283].

22. Mononitrates

22.1. Introduction

The term mononitrates refers in general to the two metabolites of ISDN, namely, IS-5-MN and IS-2-MN. These metabolites are formed in the liver. The process of denitration of ISDN to these metabolites was suggested in 1972 to be associated with inefficacy of orally administered ISDN.

However, only a few years later it was demonstrated that the mononitrates have antianginal activity resembling that of the parent compound. It was concluded that much of the vascular activity of organic nitrates results from these metabolites. It was decided then that the metabolites, rather than the parent compound ISDN, can be used for therapeutic purposes. In the mid-1970s several independent investigators concluded that IS-5-MN would be preferable to IS-2-MN for clinical use. The reasons for this are discussed in detail in the section on comparative properties of mononitrates. At present IS-5-MN is the main mononitrate studied for all the therapeutic indications for which nitrates are used. It has some advantages over ISDN, mainly due to its more favorable pharmacokinetic profile. IS-5-MN shows almost complete absorption, no considerable first-pass hepatic metabolism and high bioavailability, allowing a reduction in the number of daily doses and a prolonged duration of action.

It should be remembered, however, that although the pharmacokinetic profile of IS-5-MN is superior to that of ISDN, it is not suitable for once-daily administration in the form of conventional tablets, unless a special sustained-release oral formulation is used. Several such formulations such as IS-5-MN retard capsules have been evaluated, with good results.

In this chapter oral and intravenous administration of mononitrates are discussed together to avoid repetitions in describing the unique properties of the mononitrates.

22.2. Comparative Properties of the Mononitrates

ISDN undergoes marked denitration to IS-2-MN and, via it, to IS-5-MN. The latter compound undergoes only minimal and probably negligible first-pass metabolism. This results in the apparent advantages of IS-5-MN. It is the final active compound and its pharmacokinetic profile may be predicted more accurately than that of other nitrates. Although it has a lower vascular activity than IS-2-MN, its plasma concentration is higher [284–287]. Finally, the rate of free nitrate formation is lower with IS-5-MN - an important theoretical toxicological advantage. Therefore most laboratories studying the potential therapeutic use of mononitrates concentrate on IS-5-MN.

22.3. Comparative Hemodynamic Properties of IS-5-MN and IS-2-MN

Seidel and Michel [289] compared the hemodynamic effects of IS-5-MN and IS-2-MN in 26 patients. Invasive hemodynamic measurements were performed 10, 20 and 30 min after oral administration of 10 mg of either drug or placebo. Both agents significantly lowered systolic pulmonary arterial pressure to the same extent, by 20%. Diastolic pulmonary arterial pressure fell after IS-2-MN by 27% and after IS-5-MN by 14% only. Mean pulmonary arterial pressure was reduced by 30% after IS-2-MN and by 14% after IS-5-MN. Mean pulmonary capillary wedge pressure and left ventricular end-diastolic pressure were reduced by IS-2-MN twice as much as by IS-5-MN.

Diastolic aortic pressure was reduced by 11% after IS-2-MN. IS-5-MN increased it transiently by 5% at 10 min. Thereafter the diastolic aortic pressure returned to pretreatment values. Mean aortic pressure decreased significantly by 14% 30 min after IS-2-MN and by 6% after IS-5-MN (nonsignificant). Cardiac index increased by 12% after either drug. Heart rate was not significantly altered. Total pulmonary vascular resistance was reduced by 23% after IS-2-MN and by 17% after IS-5-MN.

It was concluded that qualitatively both drugs produced hemodynamic changes similar to those of ISDN. Quantitatively the decreases in arterial pressures after IS-2-MN were similar to those produced by ISDN and twice those produced by IS-5-MN (at equal plasma concentrations of both mononitrates).

22.4. IS-5-MN in Angina Pectoris

IS-5-MN was found effective in all forms of angina pectoris. It is usually not used for relief of pain because this long-acting compound has little advantage over other nitrates for this indication. However, it was found effective for preventing anginal episodes and reducing electrocardiographic signs of ischemia in patients with chronic stable angina pectoris (in which it also increases exercise capacity), unstable angina and vasospastic angina.

Krepp [290] studied the antianginal efficacy of oral IS-5-MN in the form of retard capsules in 30 patients with chronic stable angina pectoris. The drug was effective in all patients. Efficacy was determined, both objectively and subjectively, to range from good to very good. Six hours after the drug was first given it significantly reduced the magnitude of exercise-induced ST-segment depression. This was observed also during chronic treatment and 24 h after administration. During chronic treatment the frequency of anginal symptoms and sublingual nitroglycerin consumption were significantly reduced. Thus IS-5-MN in the form of retard capsules is effective for 24 h by once-daily administration without evidence of tolerance after prolonged use. At least one trial suggests that IS-5-MN might be superior to ISDN in chronic therapy for angina pectoris. Ohlmeier et al [110] reported that the initial effect of ISDN (sustained-release formulation, 20 mg q.i.d.) on ST-segment depression and the response of pulmonary arterial pressure to exercise was attenuated after 4 weeks of treatment. In contrast, the effect of IS-5-MN, 20 mg q.i.d., was sustained throughout this period.

Eggeling et al [291] studied the effect of slow-release IS-5-MN, 50 mg once daily, in eight patients with severe coronary artery disease. Exercise stress tests were performed before treatment, on placebo and on the active drug, at 1, 2, 4, 8 and 24 h after administration. Maximal work load increased significantly at 1–8 h after administration of IS-5-MN but returned to pretreatment level after 24 h.

Akhras and Jackson [292] studied the antianginal efficacy of monotherapy with IS-5-MN in 14 patients with coronary artery disease and chronic stable angina pectoris. After a 2-week placebo period IS-5-MN was given in a single-blind fashion, titrating the dosage at 2-week intervals, from 20 mg once daily to 20 mg twice daily, 40 mg once daily and 40 mg twice daily. The drug increased the exercise capacity significantly from 319 $\pm$ 21 sec on placebo to 394 $\pm$ 26 sec on 20 mg once daily, 434 $\pm$ 27 sec on 20 mg twice daily, 445 $\pm$ 26 sec on 40 mg once daily and 465 $\pm$ 27 sec on 40 mg twice daily (increases of 24, 36, 39 and 46%, respectively). However, a significant reduction in ST-segment depression occurred only with the twice-daily regimens. The number of anginal attacks was reduced from 21.4 $\pm$ 5.7 on placebo to 11.3 $\pm$ 4.5 on 20 mg twice daily, 10.9 $\pm$ 4.6 on 40 mg once daily and 7.1 $\pm$ 3.0 on 40 mg twice daily, a reduction of 48%, 50% and 67%, respectively. Sublingual nitroglycerin consumption decreased similarly. Thus, IS-5-MN exerts a potent antianginal effect, with no evidence of early tolerance.

Bidoggia et al [293] compared IS-5-MN with placebo in 20 patients with chronic stable angina pectoris. Exercise tests were performed after placebo and after one oral dose of IS-5-MN, 20 mg. On placebo 18 of the patients suffered angina pectoris during the exercise test, compared with only three patients on IS-5-MN. At the onset of chest pain in the 18 patients on placebo total work load was 1387 $\pm$ 851 versus 3186 $\pm$ 1178 kgm in the same patients while on IS-5-MN. The rate-pressure product at onset of pain was also significantly increased with the drug: 19.386 $\pm$ 4.285 on placebo versus 24.332 $\pm$ 4.860 on IS-5-MN. This increase was due exclusively to increase in heart rate. Despite the higher rate-pressure product the ST-segment depression under IS-5-MN was of a less magnitude than under placebo, suggesting improvement in coronary blood flow to the ischemic area.

A long-term clinical trial of IS-5-MN was reported by Schiavoni et al [294]. Nineteen patients with chronic stable angina pectoris received placebo for 10 days and then IS-5-MN, 20 mg twice daily, for 60 days. All other medications were excluded, except for sublingual nitroglycerin. The patients were evaluated clinically and electrocardiographically, but exercise tests were not performed. One patient died of acute myocardial infarction. In four of the remaining 18 patients both anginal attacks and silent ischemic episodes disappeared during the 1st week of treatment. In eight patients both types of episodes disappeared in the 2nd week of treatment. Three patients showed slight improvement and three others no improvement. In the whole group the number of anginal attacks decreased from 7.15 $\pm$ 0.88 per 10 days in the washout period to 5.74 $\pm$ 0.89 per 10 days on placebo, but the difference was not statistically significant. On IS-5-MN a significant decrease in number and severity of anginal episodes was observed. The magnitude of ST-segment depression during asymptomatic ischemic episodes was also reduced with IS-5-MN.

Kamphausen [295] studied the effect of IS-5-MN retard, 50 mg once daily, in 32 patients with postinfarction angina treated for 4 weeks. The drug significantly reduced anginal frequency, sublingual nitroglycerin consumption and exercise-induced ST-segment depression. In the most severe cases, however, monotherapy was not sufficient, and combined treatment was required.

22.5. Long-Term Use of Mononitrates in Angina Pectoris

As the development of tolerance should be studied specifically for each vasodilator and as one of the important lines of development in mononitrates is sustained-release oral formulations for once-daily administration, it is important to determine whether tolerance to IS-5-MN develops during chronic treatment.

Several investigators have reported that prolonged oral treatment with IS-5-MN is associated with a sustained effect. For example, Krepp et al [290] have demonstrated this for IS-5-MN retard capsules, given once daily, in patients with chronic stable angina pectoris.

Rennhak et al [296] studied the possible development of tolerance to long-term treatment with a conventional formulation of IS-5-MN, 20 mg twice daily, in 20 patients with angina pectoris. The drug had a marked effect on exercise-induced ST-segment depression. After 4 weeks of treatment this effect, as well as the effect on blood pressure, was not significantly attenuated.

Menzel [297] studied the antianginal effect of IS-5-MN, 20 mg two to three times daily, in 15 patients with coronary artery disease and chronic stable angina pectoris treated for 1 year. The drug reduced the exercise-induced ST-segment depression by 27.6% after 3 months of treatment, by 35.9% after 6 months and by 36.8% after 12 months. The frequency of anginal episodes was also reduced throughout the whole period of treatment.

Heepe [298] studied the antianginal efficacy of IS-5-MN retard, 50 mg once daily, in 20 patients with coronary artery disease. The drug significantly reduced the exercise-induced ST-segment depression, both at initiation of treatment and after 3 months of treatment. Work capacity had increased by 36% at the end of 3 months. A 60% reduction was observed in frequency of anginal episodes and sublingual nitroglycerin consumption.

Tauchert et al [299] reported that conventional and even moderately high doses of IS-5-MN produce a sustained antianginal effect without tolerance, whereas very high doses are associated with tolerance. They studied 24 patients with angiographically confirmed coronary artery disease. Invasive hemodynamic measurements were made at rest and during exercise, and exercise capacity was measured before and during chronic oral therapy (4 weeks) with IS-5-MN 20 to 50 mg, three times daily.

After acute administration both doses of IS-5-MN reduced mean pulmonary arterial pressure at rest by 25% with 20 mg and by 29% with 50 mg and during exercise by 30% with 20 mg and by 45% with 50 mg. Cardiac output and stroke volume were reduced only at rest. During chronic treatment with 60 and 150 mg daily the additional administration of 20 or 50 mg, respectively, lowered mean pulmonary arterial pressure by 15% and 19%, respectively, at rest, and by 22% and 14% during exercise. Exercise capacity increased slightly during chronic treatment with 60 mg daily while it decreased by 25% during chronic treatment with 150 mg daily.

These findings indicate that the unloading effect of moderate doses of IS-5-MN is sustained during chronic treatment and is associated with sustained anti-ischemic activity. In contrast, a very high dose of 150 mg daily is associated with tolerance to the vascular and anti-ischemic effects.

The potential benefit of treatment with sustained release IS-5-MN was also evaluated by an invasive hemodynamic study, reported by Jansen et al [299a]. The effects of IS-5-MN, 50 mg once daily on hemodynamics and exercise tolerance were studied after the first dose and after one week of treatment in 13 patients with angiographically proven coronary artery disease and chronic stable angina pectoris.

Two hours after administration of the first dose the mean resting pulmonary arterial pressure was significantly reduced by 27% and at submaximal exercise – by 34%. At individually highest comparable work mean pulmonary arterial pressure was reduced by 28% and at maximum work load – by 21%. After one week of therapy the hemodynamic effect was largely sustained, without any significant evidence of tolerance. Work capacity increased acutely by 60% and remained at this level at 10 h after the first dose, but not at 24 h. This improvement was fully maintained during chronic therapy [299a].

Svendsen et al [300] reported that slow-release IS-5-MN, 60 mg once or twice daily, exerts an antianginal and anti-ischemic effect sustained for 2 weeks without evidence of tolerance with either dose.

In summary, IS-5-MN exerts a sustained antianginal effect throughout long periods of treatment.

22.6. Sustained Release Oral Formulation of IS-5-MN

As with ISDN, oral treatment with IS-5-MN is complicated by a problem common to the organic nitrates, namely, the need for frequent oral dosing. The half-life of commerically available IS-5-MN is only 4–6 h. Sustained-release oral formulations have been developed to overcome this problem. One of these forms is the IS-5-MN slow-release capsule. This agent has a long duration of action, making it suitable for prophylaxis of angina pectoris.

Kenedi [301] reported a placebo-controlled, double-blind, crossover, randomized study of IS-5-MN slow-release capsules in 12 patients with chronic stable angina pectoris. Doses of 25, 50 and 100 mg, once daily, were given for 1 week each. In the 7th day of each week, 8 h after the last dose, an exercise stress test was performed. The drug produced a significant (compared with placebo) and dose-dependent decrease in the magnitude of exercise-induced ST-segment depression: 34.4% with 25 mg, 47.6% with 50 mg and 65.7% with 100 mg. The frequency of anginal episodes and nitroglycerin consumption was also significantly reduced. Thadani et al [302] evaluated the effect of another slow-release formulation of IS-5-MN, 50 and 100 mg, compared with placebo, on exercise tolerance in nine patients with stable angina pectoris. Exercise time increased by 62 sec at 4 h ($p < 0.05$) and by 7 sec at 20 h with the 50-mg dose compared to placebo. After the 100 mg dose exercise time increased by 87 sec ($p < 0.05$) at 4 h and by 11 sec at 20 h. Both doses were ineffective at 24 h, and the effect at 20 h was nonsignificant. The investigators suggested that this time course of effect results from rapid development of tolerance as plasma levels of the drug were still high at 24 h despite the lack of effect. This explanation is, however, unlikely, as another report from Germany revealed that a conventional formulation of IS-5-MN, 20 mg twice daily, does not cause tolerance [311]. The pharmacokinetic difference is not so marked to account for difference in efficacy.

Wisenberg et al [311a] studied the antianginal and anti-ischemic effects of a single daily dose of IS-5-MN, 60 mg slow-release formulation, in 18 patients with chronic stable angina pectoris. The slow release formulation was compared with a high-dose ISDN, conventional formulation, 30 mg 4 times daily, in a double-blind, randomized placebo-controlled cross-over study. Exercise tests were performed on the first day of therapy and after 11 to 14 days of continued therapy. On the first day both drugs produced a significant prolongation of exercise duration to moderate angina, in comparison with placebo. ISMN, taken in the morning, increased the exercise duration by 23% at 12:30 p.m., 19% at 5:00 p.m., and 13% at 08:30 p.m. For ISDN the respective values were 19%, 24% and 21%. There were no significant differences between the effects of both drugs.

After 11 to 14 days of continuous treatment only slow-release IS-5-MN retained a significant beneficial effect with prolongation of exercise duration by 21% at 12:30 p.m. and 14% at 5:00 p.m., but no effect at 8:30 p.m. For this drug the plasma levels paralleled the improvement in exercise tolerance. The sustained effect can probably be attributed to the decrease in plasma levels into the subtherapeutic range [311a].

Eggeling et al [304] studied the effect of slow-release IS-5-MN (pellet formula) in ten patients with angiographically confirmed two- and three-vessel coronary artery disease. Bicycle exercise tests in supine position were performed, with a starting work load of 25 w and increments of 25 w every 2 min. Tests were performed without drugs, after placebo and at 1, 4, 8 and 24 h after intake of 50 mg IS-5-MN. Maximal working capacity increased at 1, 4 and 8 h after medication. ST-segment depression was reduced at 1, 4 and 8 h. At 24 h no alterations in working capacity and ST-segment depression could be detected. Heart rate and arterial pressure were not significantly altered. Other studies of sustained-release IS-5-MN are discussed in the section on IS-5-MN in angina pectoris.

22.7. *Mononitrates in Coronary Spasm and Variant Angina*

The role of nitrates in prevention of episodes of coronary spasm in unstable angina has been described earlier in this book. Because long-term intravenous treatment is not practical, and because IS-5-MN might have pharmacokinetic properties superior to those of ISDN, this agent has also been studied recently in variant angina.

Distante et al [89] evaluated the effectiveness of oral IS-5-MN in preventing episodes of myocardial ischemia induced by provocative tests such as ergonovine and isometric exercise test (handgrip). Three patients had positive ergonovine tests with chest pain, ECG changes, echocardiographically demonstrated myocardial hypokinesia and angiographically demonstrated coronary spasm. The study was repeated in all three patients 1 h after oral administration of IS-5-MN (20–40 mg), but no marker of myocardial ischemia was observed, even when the dose of ergonovine was doubled. The drug prevented myocardial ischemia also in four patients in whom isometric exercise test with placebo had produced ischemia.

In addition to suppression of vasospasm the drug had also beneficially affected the determinants of myocardial oxygen demand (reduction of double-product).

22.8. *Effects of IS-5-MN on Cardiac Function and Hemodynamics in Patients with Coronary Artery Disease*

The cardiovascular and hemodynamic effects of IS-5-MN in patients with coronary artery disease resemble those of other nitrates. Under resting condition the major effect is reduction in left ventricular filling pressure, mainly due to the venodilatory effect of the drug. Cardiac output is not increased, and systemic vascular resistance is not decreased, as expected in patients without heart failure.

Biamino et al [305] reported an invasive hemodynamic study of the effect of a single oral dose of IS-5-MN, at rest and during exercise, in ten patients with coronary artery disease and chronic stable angina pectoris. Mean pulmonary arterial pressure fell from 16.1 ± 2.3 to 12 ± 2.6 mmHg and right atrial pressure from 8.4 ± 2.3 to 4.9 ± 2.6 mmHg. These changes, which were highly significant, reached their peak values 45 min after oral administration of the drug. Right ventricular filling pressure, cardiac output and systemic vascular resistance were not significantly altered.

These hemodynamic changes were accompanied by elevation of threshold for angina pectoris under stress conditions. This was observed in nine of ten patients studied. Left and right filling pressures were significantly reduced during exercise. The pulmonary capillary wedge pressure was reduced from 31 ± 9.5 to 16 ± 7.1 mmHg.

Brüggemann et al [306] reported that in patients with coronary artery disease IS-5-MN, 40 mg, did not alter ejection fraction at rest but increased it by 16% during exercise.

Bergbauer and Sabin [307] evaluated the effects of IS-5-MN on left ventricular function by systolic time intervals in 25 patients with angiographically confirmed two- or three-vessel coronary artery disease. All patients had elevated pulmonary arterial pressure at rest and during exercise. After 20 mg IS-5-MN there was a significant reduction in the preejection period and prolongation of left ventricular ejection time. The ratio of preejection period/left ventricular ejection time was reduced from a mean of 0.31 to 0.20. Left ventricular diastolic dimension fell by 10% and systolic dimension by 17%. The shortening fraction increased from 24% to 31%. Thus acute administration of 20 mg IS-5-MN produced a significant improvement of the echocardiographically determined parameters of left ventricular function.

The response of heart rate to IS-5-MN was variable. In most studies heart rate was not altered or was slightly increased. In some cases, however, (mainly those with congestive heart failure) decreases in heart rate have been observed. The possible development of hemodynamic tolerance was specifically evaluated for IS-5-MN. At low doses this agent did not produce hemodynamic tolerance. Müller et al [308] found no decrease in the efficacy of the drug after 2 weeks of treatment with 20 mg t.i.d.

22.9. Comparative Studies of Mononitrates and Other Drugs in Angina Pectoris

22.9.1. Calcium Antagonists

As mononitrates are relatively new agents, comparative studies with other antianginal drugs have only recently been performed. The first attempts were, of course, to compare mononitrates with calcium antagonists.

Both calcium antagonists and nitrates reduce preload and afterload, thereby reducing myocardial oxygen consumption. However, calcium antagonists have a more pronounced effect on afterload and nitrates on preload. This is particularly true for IS-5-MN, which was found in preliminary studies to exert a greater effect on left ventricular filling pressure and less effect on systemic vascular resistance, compared with the effect of nitroglycerin.

It should also be remembered that calcium antagonists have a negative inotropic effect which might also contribute to the reduction in oxygen consumption. Koch et al [309] compared the efficacy of one dose of IS-5-MN (20 mg) with that of two doses of nifedipine (5 and 20 mg), using a double-blind, placebo-controlled, randomized, crossover design in 21 patients with exercise-induced angina pectoris. Nifedipine increased heart rate at rest whereas IS-5-MN did not alter it. Nifedipine lowered systolic blood pressure during exercise more than IS-5-MN. However, both agents had a comparable hypotensive effect at rest. Both doses of nifedipine and IS-5-MN lowered the double product to the same extent. Total duration of exercise, total work performed and maximal workload increased significantly only after IS-5-MN. The exercise until occurrence of ST-segment depression of 1 mm or more increased significantly with IS-5-MN as compared with placebo and nifedipine. A significant decrease in the exercise-induced ST-segment depression occurred more frequently with IS-5-MN than with nifedipine. It was concluded that the antianginal efficacy of IS-5-MN is equal or even superior to that of nifedipine, 5 to 20 mg.

It should be remembered that the results of this acute study can not be extrapolated to chronic therapy. Further long-term studies are required. One such study was recently reported by Erlemeier et al [310] who compared nifedipine retard, 20 mg twice daily, slow-release IS-5-MN, 50 mg once daily, and their combination in nine patients with chronic stable angina pectoris, treated for 2 weeks. All three forms of treatment significantly reduced the frequency of anginal episodes and the magnitude of ST-segment depression. The effect of combined therapy was better than that of monotherapy. However, subjective evaluation of the drugs rated the combination worst, perhaps due to orthostatic hypotension.

22.9.2. Other Nitrates

IS-5-MN has been compared with ISDN in large clinical studies not involving an exercise test. Almost 600 patients, treated outside hospital, were studied. The dose of IS-5-MN could be reduced compared with conventional doses of slow-release ISDN, and good antianginal efficacy was observed [311, 312].

Steinorth et al [313] reported a double-blind, randomized, crossover comparative study of slow-release ISDN, 80 or 120 mg daily, and IS-5-MN, 40 or 60 mg daily. Each drug was given for 4 weeks to 90 patients. Sublingual nitroglycerin consumption and the frequency of anginal episodes were comparably reduced by 40 mg IS-5-MN and 80 mg ISDN slow-release, as well as by 60 mg IS-5-MN and 120 mg ISDN slow-release. There was, however, an improvement in symptoms of angina pectoris when ISDN was replaced by IS-5-MN. Thus a clinical superiority of the mononitrate was observed although its dose was only half that of the dinitrate.

The two drugs had similar adverse effects and tolerability. These findings indicate that IS-5-MN may replace ISDN for the long-term prophylaxis of angina pectoris. Tauchert et al [314] studied the hemodynamic and exercise parameters of IS-5-MN compared with placebo and transdermal nitroglycerin patches in 47 patients with coronary artery disease. A single oral dose of IS-5-MN, 20 mg, increased work capacity from 268 ± 170 to 432 ± 264 w/min. A comparable effect was achieved only by four patches of transdermal nitroglycerin (each delivering 5 mg/24 h). These findings suggest that oral administration of conventional formulations of IS-5-MN may be superior to low doses of transdermal nitroglycerin. It should be remembered, however, that patches of 5 mg/24 h are now uncommonly used, and that comparative studies with higher doses should be performed.

22.10. Mononitrates in Congestive Heart Failure

In congestive heart failure mononitrates exert the usual effects of nitrates, the predominant of which being a decrease in left and right filling pressures, associated with a decrease in systemic vascular resistance and an increase in cardiac output.

Rabinowitz et al [315] studied the hemodynamic effects of intravenous IS-5-MN in 17 patients with congestive heart failure complicating an acute (ten cases) or old (seven cases) myocardial infarction. The dose of IS-5-MN was titrated to achieve optimal effect. At the maximal or optimal dose mean pulmonary capillary wedge pressure decreased from 25.4 ± 4.6 to 18.3 ± 5.9 mmHg, right atrial pressure decreased from 9.1 ± 5.6 to 6.3 ± 4.2 mmHg, cardiac index increased from 2.08 ± 0.4 to 2.3 ± 0.3 L min^{-1} m^{-2}, and systemic vascular resistance decreased from 1.778 ± 426 to 1.610 ± 298 dyn sec cm^{-5}. The duration of action of IS-5-MN on left ventricular filling pressure was longer in patients with acute than in those with chronic congestive heart failure. Chronic patients also showed a greater response of systemic vascular resistance. Five patients, who were previously unresponsive to other vasodilators, showed a beneficial response to IS-5-MN. Without more data, it is difficult to conclude whether this resulted from the properties of IS-5-MN or from differences in dosing.

Good results were also observed with orally administered IS-5-MN. Schiavoni et al [316] studied the hemodynamic effects of a single oral dose of 40 mg IS-5-MN for 6 h after administration in 15 patients with chronic congestive heart failure. Right atrial pressure was reduced by 40% between 120–140 min after administration, pulmonary systolic arterial pressure decreased by 25%, and left ventricular filling pressure by 30%. Systolic blood pressure was significantly decreased. Heart rate and cardiac output were not significantly altered. Klütsch et al [317] reported improvement of noninvasive indices of cardiac function after prolonged oral treatment with IS-5-MN in patients with chronic congestive heart failure.

In conclusion, IS-5-MN administered for short periods, intravenously or orally, appears to have a beneficial role in the treatment of acute as well as chronic congestive heart failure.

Maintenance therapy by IS-5-MN deserves further long-term evaluation. This has been recently reported by Tronconi and coworkers [318] using oral, chronic IS-5-MN. These investigators studied 20 patients with congestive heart failure due to coronary artery disease. They were treated with 20 mg oral IS-5-MN daily for 23.9 ± 4.5 days and evaluated noninvasively. A long-term improvement in left ventricular performance and symptoms was demonstrated. This included an improvement in exercise duration and workload, a slight reduction of left ventricular dimension, and an improvement of echocardiographic indices of left ventricular dysfunction. The main problem with this study is, of course, lack of long-term invasive hemodynamic evaluation.

As we know from experience with other vasodilators, this is clearly required. A recent invasive hemodynamic study with Elantan 20 confirmed that long-term treatment with IS-5-MN is not associated with tolerance in patients with congestive heart failure. Keck [319] studied 20 patients with chronic congestive heart failure, New York Heart Association class II–III, by cardiac catheterization and 2-dimensional echocardiography. IS-5-MN was given at a dose of 20 mg three times daily. After administration of the first dose the mean pulmonary capillary wedge pressure decreased by 17%, left ventricular end-diastolic diameter by 3% and end-systolic diameter by 7.7%. During chronic treatment a single oral dose of 20 mg produced a comparable effect: decreases of 18%, 4.5% and 11%, respectively. Thus, no tolerance to the main hemodynamic effects of IS-5-MN has been observed during chronic therapy in patients with congestive heart failure. VandenBurg et al [320] performed repeated cardiac catheterizations in six patients with chronic congestive heart failure treated with diuretics. They received IS-5-MN orally for 4 weeks. Hemodynamic measurements were performed after acute therapy and at the end of 4 weeks of chronic therapy. Acute administration of IS-5-MN reduced pulmonary arterial diastolic pressure by 12.5 ± 2.8 mmHg, pulmonary arterial systolic pressure by 19.2 ± 14.6 mmHg, mean right atrial pressure by 2.8 ± 2.5 mmHg and systemic systolic arterial pressure by 11.7 ± 10.8 mmHg ($p <$ 0.05, in all cases). The other hemodynamic parameters were not altered. After chronic therapy pulmonary diastolic and systolic pressures were significantly lower than after acute treatment.

These results suggest that IS-5-MN may be a suitable long-term therapy for patients with congestive heart failure.

Although IS-5-MN was investigated mainly in chronic heart failure (as its pharmacokinetic profile is superior to that of other nitrates for long-term therapy), it was found effective also in acute heart failure. This was demonstrated in the study by Rabinowitz et al [315], discussed earlier in this chapter.

Recently Gammage et al [321] studied the hemodynamic response to an intravenous bolus of IS-5-MN in 11 patients with acute left ventricular failure complicating acute myocardial infarction. All patients had clinical and radiographic findings of congestive heart failure, confirmed by pulmonary capillary wedge pressure of more than 20 mmHg. Three bolus injections of IS-5-MN, 5 mg, at 15-min interval, were given intravenously. At 45 min the mean pulmonary capillary wedge pressure fell from 26.2 ± 6.8 mmHg to 17.5 ± 8.5 mmHg without significant changes in systemic arterial pressure, stroke volume index, stroke work index and systemic vascular resistance. Heart rate decreased from 103 ± 11 to 97 ± 15 beats/min, and cardiac index from 2.4 ± 0.8 to 2.3 ± 0.7 $min^{-1}m^{-2}$. All patients had also a rapid symptomatic improvement. These findings indicate that intravenous bolus of IS-5-MN can be effectively and safely used in acute left ventricular failure.

In summary, mononitrates produce hemodynamic improvement, resulting mainly from venodilation, in patients with acute and chronic congestive heart failure. The effect is sustained throughout prolonged treatment periods and is associated with symptomatic improvement.

Recently it has also been shown that long-term treatment with IS-5-MN increases exercise capacity in patients with congestive heart failure. Pozzoli et al [322] studied ten patients with congestive heart failure, functional class II, who after 3 weeks of washout received IS-5-MN, 20 mg twice daily for 3 weeks. Bicycle exercise tests were performed before treatment, at the end of 3 weeks of treatment and 1 week later. The drug significantly increased exercise capacity.

22.11. IS-5-MN Compared with ISDN in Heart Failure

Rabinowitz [323] compared the hemodynamic, clinical and metabolic effects of IS-5-MN and ISDN given at equal doses in a randomized, crossover study of 20 patients with congestive heart failure due to acute (1-week) or old myocardial infarction. Invasive hemodynamic measurements were made before and after oral administration of a single dose of 40 mg of either drug. The second drug was given after the hemodynamic parameters reverted to pretreatment levels. The drug produced qualitatively similar effects, the predominant of which being a reduction in left and right filling pressures. Systemic vascular resistance decreased and cardiac output increased. The maximum effect was at 1–2 h for both drugs. However, the effect of IS-5-MN was maintained for up to at least 4 h, while that of ISDN disappeared at this time.

It may be summarized that, at equal oral doses, IS-5-MN is equipotent to ISDN, with a somewhat longer duration of action of IS-5-MN.

Tanaka et al [324] compared the hemodynamic effects of orally administered IS-5-MN, 40–80 mg, and ISDN, 10 mg, in 14 patients with chronic congestive heart failure. IS-5-MN significantly lowered the pulmonary capillary wedge pressure from 20.1 ± 2.7 to 13.4 ± 3.7 mmHg and right atrial pressure from 9.6 ± 5.0 to 6.5 ± 4.0 mmHg, while ISDN reduced these parameters from 20.6 ± 2.3 to 15.0 ± 4.4 mmHg and from 9.6 ± 2.6 to 6.4 ± 2.4 mmHg. However, the effect of IS-5-MN was more long-lasting than that of ISDN (6 h versus 2 h). These findings suggest that IS-5-MN might be of marked benefit, more than ISDN, in patients with congestive heart failure in whom continued preload reduction is necessary.

22.12. IS-5-MN in Acute Myocardial Infarction

IS-5-MN may be superior to other nitrates in patients with acute myocardial infarction. This is due to the findings (which are as yet preliminary) that IS-5-MN reduces left ventricular filling pressure to a greater extent than nitroglycerin, in the presence of less reduction in systemic arterial pressure. Therefore it is important to evaluate the effects of IS-5-MN in patients with acute myocardial infarction.

Bennett et al [324a] evaluated data from 250 patients with acute myocardial infarction who entered a multicenter, double-blind, randomized, placebo-controlled, closed sequential trial of 5-day treatment with IS-5-MN. The first dose was administered within 6.2 ± 4.6 h from onset of pain in the IS-5-MN group and 7.6 ± 8.1 h in the placebo group. The 5-day mortality in the two groups was not significantly different. In the first 48 h fewer patients in the IS-5-MN group required morphine (32% versus 40%) or long-acting nitrates (17% versus 28%). Severe hypotension developed in 6% of the patients who received IS-5-MN and 9% of those receiving placebo. Headache was more common in the patients receiving IS-5-MN.

The investigators concluded that these preliminary data indicate a potential beneficial effect of IS-5-MN in patients with acute myocardial infarction. Demonstration of significant difference in mortality between the two groups requires the study of a larger number of patients.

23. Transdermal Administration of Nitrates

Nitrates are given transdermally to avoid first-pass hepatic metabolism and to achieve constantly high plasma concentrations. The most widely used method is transdermal nitroglycerin patches (discs). Nitroglycerin and ISDN ointments are also available.

This chapter summarizes mainly the experience with transdermal nitroglycerin discs. Other formulations are briefly mentioned, as the experience with them is limited.

23.1. Transdermal Nitroglycerin Discs

The realization that nitroglycerin can be delivered transdermally, that this method has pharmacokinetic advantages, and that the early delivery systems such as the nitroglycerin ointment are impractical has led to the development of a standard system – the transdermal nitroglycerin disc. Since 1982 several such discs, developed independently by various pharmaceutical companies, have been marketed in the United States. The short history of these discs has been described by Abrams as "one of the most remarkable pharmaceutical stories [330]."

The first generation transdermal nitroglycerin discs achieve a relatively constant plasma level of the active drug throughout 24 h of treatment – the usually recommended application time. This is achieved by a constant nitroglycerin release (about 0.5–0.6 mg cm^{-2} 24 h^{-1}) from the disc on its side applied to the skin. This method has several advantages:

1. Nitroglycerin can be used not only over short periods for relief of pain in patients with angina pectoris but also over long periods and for prevention of pain.
2. Nitroglycerin may be used for long periods in new indications such as chronic congestive heart failure.
3. The once-daily administration simplifies the therapeutic regimen by eliminating the need for frequent dosing.
4. The first-pass hepatic degradation of orally administered nitrates is avoided.
5. The discs obviate the inconvenience of nitrate ointment.
6. Relatively constant plasma levels are maintained. (This can be considered also as a disadvantage).

The nitroglycerin discs have been very well accepted, both by patients and physicians. Some investigators have suggested that the long-term use of discs is associated, perhaps more than other delivery systems, with problems of nitrate therapy such as tolerance. The efficacy of the discs has also been questioned. These points have been carefully studied in the past 3 years. It may be said that although the story is not yet complete, as claim some of the opponents of this therapeutic system, wise use of the discs gives the clinician an important new therapeutic tool which can substantially improve clinical practice without the addition of new risks.

It should be remembered that other organic nitrates can also be administered transdermally, and that ointments of these nitrates are available. The reason that they were not chosen for the discs is that the special properties of the organic nitrates, and mainly the long duration of action, are not required when 24-h delivery systems are used. Moreover, nitroglycerin is excellent for transdermal delivery because it has a low molecular weight and high lipophilicity.

Nitroglycerin is not the only drug which is routinely given transdermally in clinical practice. In 1980 a transdermal device containing scopolamine was introduced to clinical use. It provides controlled delivery of scopolamine over 3 days and is effectively used for treatment of motion sickness. A transdermal disc of clonidine is available for clinical use in the United States. It provides controlled delivery of clonidine over a 7-day wearing period and is effective in treatment of hypertension. A transdermal delivery system for estradiol for use in the postmenopausal syndrome is now being developed. Thus, transdermal nitrates are part of a new and important field in drug delivery systems. As the potential of rate-controlled delivery is greatest for potent drugs with a low therapeutic index and short elimination half-life, nitroglycerin is near-optimal for this method.

Nitroglycerin discs have been used mainly for treatment of chronic stable angina pectoris and chronic congestive heart failure. There is also some experience in their use for treatment of unstable angina and for congestive heart failure after acute myocardial infarction.

23.2. Comparative Efficacy

Comparative evaluation of the various nitroglycerin discs available can be made by specifically designed comparative studies or by comparing data from different studies. Comparative studies, if performed, have not been published. Such studies are difficult to perform and would require large numbers of patients to undergo repeated exercise tests (in the case of angina pectoris) or prolonged invasive hemodynamic monitoring (in the case of congestive heart failure). It is unlikely that such studies will be reported in the near future.

In comparing different reports of transdermal nitroglycerin discs the main problem is to determine nitroglycerin dose equivalency between studies [115, 325–329].

It is still controversial which method is superior in comparing discs: dosage determined by the amount of drug delivered over 24 h [330] or the plasma concentrations of nitroglycerin [331]. In many studies neither method is used. In my opinion and in view of the lack of comparative data a decision should be made based on efficacy studies. The clinician should chose for each indication that disc which "scores" high in efficacy studies. Perhaps studies with other discs, performed with higher doses, different methods, etc. will in the future show the superiority of another system. Until then the clinician must rely upon those efficacy studies that he reads. For the convenience of the clinician several such studies are discussed in this chapter under the different indications for treatment.

23.3. Nitroglycerin Discs in Angina Pectoris

Because of its short elimination time sublingually administered nitroglycerin has until recently been limited to the relief of pain in patients with chronic stable angina pectoris. Studies of continuous intravenous infusion of nitroglycerin have revealed that constantly high plasma levels of the drug can be effective not only for the relief of pain but also for the prevention of its development.

As intravenous therapy is limited to hospitalized patients, the development of transdermal nitroglycerin discs has created the first practical basis for keeping constant plasma levels of the drug in ambulatory patients and for its use for antianginal prophylaxis, namely, the prevention of development of anginal episodes.

Several studies have revealed the efficacy of transdermal nitroglycerin discs in antianginal prophylaxis. The usual effect is a 30% increase in exercise duration soon after application of the disc and a somewhat smaller effect at 24 h and thereafter (15% increase). Anginal frequency and sublingual nitroglycerin consumption are usually reduced by 40%–60%.

Greco and coworkers [332] recently studied in a double-blind, randomized, crossover trial the antianginal and anti-ischemic efficacy of a transdermal nitroglycerin disc. This disc, releasing 10 mg nitroglycerin over 24 h, was assessed in 19 outpatients with chronic stable exercise-induced angina pectoris. The trial consisted of a 3-day washout, a first 1-week period with nitroglycerin or placebo patches and a second 1-week period with the other patch in trial. No other antianginal therapy except for sublingual nitroglycerin was allowed.

Treadmill exercise tests were carried out at the end of washout before patch application and after 7 days of application of nitroglycerin or placebo patches. Seventeen patients completed the trial. The frequency of anginal episodes was reduced by 31.3% and sublingual nitroglycerin consumption by 34.3% ($p < 0.01$) during the week with nitroglycerin compared with placebo. Exercise duration increased by 29% ($p < 0.001$) and 16.1% ($p < 0.01$) at 3 and 24 h, respectively, with nitroglycerin compared to placebo. ST-segment depression at comparable work load decreased by 69% and 40.5% ($p < 0.01$) at 3 and 24 h, respectively, with nitroglycerin patch compared to placebo. Onset of angina was delayed and maximum rate-pressure product significantly increased at 3 and 24 h of treatment.

Based on these data it may be concluded that nitroglycerin patches exert antianginal and anti-ischemic effects for up to 24 h after application. Although the observed effects, and especially those on exercise duration, are only modest, it should be remembered that the dose used is within the range of low-dose therapy [330]. If higher doses had been used, a better response might be expected.

Perhaps the most important finding of this study is the demonstration of efficacy for 24 h. It should be remembered that in a review reported late in 1984 it was stated that "except for one study, no studies have been reported that indicate that transdermal discs of nitroglycerin at any dosage provide anti-anginal protection for 24 hours" [330]. Greco and coworkers [332] have now clearly shown a 24-h efficacy for the nitroglycerin patch.

Rezakovic et al [333, 333a] studied the chronic efficacy of the nitroglycerin patch 5 mg/24 h, in two trials in patients with chronic stable angina pectoris. The first study was a randomized, double-blind, placebo-controlled, crossover study in 40 patients, who received the drug and placebo for 2 weeks each. The second was an evaluation of 12 months of nitroglycerin patch in 24 patients with washout periods at the beginning and at the end of the study, as well as with a single-blind 7-day placebo period at the end of the study. Exercise tests were performed after 2 weeks and 3, 6, 9 and 12 months of treatment, 5 h after application of the patch. An additional exercise test was performed 24 h after dosing, after 12 months of treatment.

At 14 days the nitroglycerin patch decreased by 61% the number of episodes of ST-segment depression of 1.5 mm or more. This was associated with a decrease of 39% in the number of premature ventricular beats. Nitroglycerin patch, compared with placebo, prolonged exercise time to onset of 1-mm ST-segment depression and chest pain and augmented the maximum walking time with diminished maximum ST-segment depression and faster recovery of depressed ST-segment in the 3rd and 6th min after exercise test in both trials. All these results were significantly better than those achieved with placebo, at 5 h as well as 24 h after application and at 14 days as well as 12 months of treatment. It may be concluded that the transdermal nitroglycerin patch has a potent and sustained anti-ischemic therapy for at least 12 months.

Schiavoni et al [334] studied the antianginal efficacy of monotherapy with nitroglycerin percutaneons, 10 mg/24 h, in 72 patients. The drug increased exercise duration in exercise tests performed at 6 and 24 h after administration by 29% and 28%, respectively. At 28 h exercise duration was not significantly increased. Total work increased significantly by 41% and 39% at 6 and 24 h, and nonsignificantly, by 4% only, after 28 h.

Good results were achieved also in long-term studies. (In this field "long-term" studies are those in which treatment is continued for at least 1 week. Longer studies are definitely required.) Georgopoulos et al [335] reported a double-blind, placebo-controlled, crossover trial of transdermal nitroglycerin 5 mg/24 h, in 13 patients treated for 2 weeks. During the treatment period ST-segment depression was reduced by 51% compared with placebo, at 2 h and 24 h. Sublingual nitroglycerin consumption was reduced by 60%. Thompson [336], presenting data from a New Zealand study, reported that transdermal nitroglycerin discs increased exercise duration by 30% at 2 h and by 25% at 26 h. Nitroglycerin consumption was reduced by about 60%.

Greco et al [336a] studied the chronic efficacy of 10 mg transdermal nitroglycerin patch in 30 patients with stable angina pectoris. After a 7-days washout period the patients entered a 2-months randomized double-blind placebo-controlled cross-over phase. The drug improved exercise duration to angina and ischemia and shortened the period of post exercise recovery of pain and ST-segment depression. The effects on time to angina and ischemia persisted during chronic therapy but the effects on recovery were partially attenuated.

Atkins [337] has recently reviewed several studies of transdermal nitroglycerin which used exercise time to evaluate efficacy. When only results of the first patch, at initiation of therapy were analyzed, the duration of action of nitroglycerin patches was 20–26 h. It is also clear that therapeutic plasma levels of nitroglycerin are maintained for 24 h or even longer after administration of the patch [338].

Several studies demonstrated comparable antianginal efficacy of transdermal nitroglycerin and oral ISDN. Imhof [339] compared the effect of ISDN, 80 mg daily, and transdermal nitroglycerin, 5 mg/24 h, in a double-blind, randomized, crossover trial. The drugs increased exercise time by 34% and 36%, respectively. The difference was not significant. Pucci et al [340] compared the effect of long-acting ISDN, 20 mg three times daily, and transdermal nitroglycerin, 10 mg/24 h, in ten patients. Both drugs increased exercise duration to the same extent. Total work increased by 64% at 3 h and 45% at 24 h after acute administration of the drugs. Similar effects of nifedipine, 20 mg twice daily and transdermal nitroglycerin, 5 mg/24 h, were observed by Wester et al [341].

Fazio et al [342] studied the effects of transdermal nitroglycerin on physical work capacity and regional myocardial perfusion as determined by thallium-201 stress scintigraphy. Ten patients with exercise-induced chronic stable angina pectoris underwent an exercise test until the onset of chest pain, 20 h after administration (at different days) of placebo or 10 mg transdermal nitroglycerin. Scintigraphic studies were performed 10–15 min after stress test and 3–4 h after reperfusion. Transdermal nitroglycerin significantly increased the work capacity, by 64%, and the myocardial perfusion during exercise, as determined by scintigraphy.

Some investigators, studying other transdermal nitroglycerin discs, revealed less favorable results. For example, Reichek et al [328], using another disc, reported that low-dose therapy (mean dose, 9.3 mg per 24 h) exerts only a small antianginal effect. At higher doses a more considerable effect was observed, but for less than 24 h. An important weakness of this study is the small number of patients. However, even if one accepts these results, in my opinion they might indicate more a problem with the specific device than with the method. It should be remembered that in the view of some investigators transdermal therapy should not be over 24 h but rather take about 16 h, with interruption during the night, as a method to prevent development of tolerance.

James et al [343] studied the antianginal efficacy of transdermal nitroglycerin 5 mg/24 h, in 12 patients. The drug increased exercise time to angina and to 1-mm ST-segment depression and total exercise time at 3 h after administration but not at 24, 48 and 72 h.

Hollenberg and Go [325] reported that transdermally administered nitroglycerin at a dose range of 5–20 mg per 24 h, markedly and significantly reduced the exercise-induced ST-segment depression, by 31%, in ten patients with stable angina pectoris. Exercise time to 1-mm ST-segment depression increased by 33%. However, these investigators revealed only a small prolongation of duration of exercise to angina, even with high doses of nitroglycerin. This was the case in both acute and long-term studies. Crean et al [326] studied transdermal nitroglycerin, 5 mg/24 h, in 11 patients, treated for 1 week with the drug and another week with placebo. The drug did not prolong exercise duration compared with placebo at 4 h after administration. The total duration of ST-segment depression, assessed by continuous ECG monitoring, was slightly but nonsignificantly reduced. Hollenberg and Go [325] evaluated the effect of transdermal nitroglycerin in five patients with chronic stable angina pectoris treated for 4 weeks. The decrease in systolic blood pressure and the associated increase in heart rate produced by a single patch were significantly smaller during the 2nd and 4th weeks of therapy than at initiation of therapy. However, the anti-ischemic effects of transdermal (and of sublingual) nitroglycerin were maintained despite attenuation of the peripheral effect.

It is possible that objective evaluation of the effect on exercise duration may miss subjective effects. Muiesan et al [344] reported that in 13 patients transdermal nitroglycerin had no effect on exercise duration but produced a 42% reduction in frequency of angina. In most reported studies of transdermal nitroglycerin anginal frequency was reduced by 40%–60%.

Charash and Scheidt [345], who reviewed in 1986 the studies with transdermal nitroglycerin, stated that the results of acute as well as long-term studies vary due to differences in study design, small series, interpatient physiologic variability and the use of fixed rather than titrated doses. I believe that equally important are the differences between the discs of different manufacturers.

It has been previously suggested that the antianginal effect exerted by transdermal nitroglycerin discs is less than that exerted by sublingual nitroglycerin tablets in the same patients [328, 325]. This finding, however, not only suffers from all the problems of comparative studies with different delivery systems but also from the expected difference between acute and long-term nitrate therapy. Moreover, even patients treated with transdermal nitroglycerin discs or long-acting nitrates can acutely take nitroglycerin oral spray or sublingual tablets of nitroglycerin prior to exceptionally heavy work loads. It should also be remembered that even if acute administration of a nitroglycerin tablet does have a greater antianginal potency than that of transdermal nitroglycerin discs, it is not the purpose of the discs to overcome the efficacy of sublingual nitroglycerin acutely. The advantages of the discs are in long-term treatment, prevention and improvement of compliance by simplification of complex regimens.

Another controversy concerns the dose of transdermal nitroglycerin. Several investigators have tried to improve efficacy or to overcome attenuation of the effect by increasing the doses of transdermal nitroglycerin to 40 mg daily or, occasionally, even more. These high doses, resulting in constantly high plasma levels, might contribute to the development of tolerance. I believe that the dose should be as low as possible. Greco and coworkers [332] have shown that the antianginal effect of a low ("conventional") dose of transdermal nitroglycerin, 10 mg/24 h, exerts a sustained antianginal effect.

Most studies have used doses of 5–45 mg/24 h. The number of acute studies in which the drug was effective is comparable with low and high doses. In chronic studies there is some tendency towards better efficacy of higher doses. I believe, however, that low doses are quite effective and should be preferably used, mainly to prevent tolerance. It is the experience of my colleagues and myself, working with nitroglycerin patches, that a dose of 10 mg/24 h, is sufficient for most patients, 5 mg/24 h are probably not enough. This is supported by several studies of other investigators.

Martines et al [346] performed a double-blind, randomized, crossover study of nitroglycerin discs, 5 and 10 mg/24 h as monotherapy, and placebo in patients with angina pectoris. A statistically significant increase in exercise duration was observed with 5 mg/24 h compared with placebo. A further significant increase was observed with 10 mg/24 h compared with placebo and 5 mg/24 h. This confirms that a dose of 5 mg does not achieve the maximal benefit of this form of therapy, and that a 10-mg dose is probably required. Similar findings were reported by Schneidet et al [347] for doses of 2.5, 5 and 10 mg/24 h.

Parker and Fung [115], however, reported that high doses, up to 45 mg/24 h, resulted in a greater increase of exercise time than doses of 15 and 30 mg. The duration of action of the higher doses was longer than that of the lower doses.

Until further studies on high-dose nitroglycerin discs are available, it is recommended to titrate the doses individually but to try to stay within the low-dose range, preferably 10 mg. The dose-response curve of transdermal nitroglycerin is relatively flat, and the slight additional benefit that might be achieved by higher doses is counteracted by the problems associated with long-term use of such doses.

It has been stated that the transdermal disc might have a particularly prominent placebo effect in treatment of angina pectoris. However, at present there are several placebo-controlled studies which show superiority of the verum over placebo. A large-scale placebo-controlled study is now being performed in Israel.

In any case it is agreed that transdermal nitroglycerin patches are very well accepted by the patients. (At least patients compliance is not a potential problem.)

Rezakovic-Music [348] has recently reviewed the literature on transdermal nitroglycerin patches and pooled the data. Although this approach has considerable limitations, it allows some perspective in this extensively investigated field.

The following results were found:

1. A significant anti-ischemic and antianginal efficacy of nitroglycerin patches in angina pectoris was demonstrated in 43 acute studies with 618 patients and in 33 chronic studies including 669 patients. No significant therapeutic effects were observed in four acute studies with 56 patients and in 16 chronic studies with 226 patients.
2. A 24-h duration of efficacy was demonstrated in 20 acute trials including 353 patients and in 16 chronic studies including 276 patients. The effect was sustained throughout a follow-up period ranging from several days to 12 months of continuous patch application.

3. A 24-h duration of antianginal activity after acute patch administration was demonstrated in 67% of patients. In 26% of patients the efficacy was attenuated after 12 h, and in 7% no effect was observed.
4. A 24-h efficacy during chronic therapy was demonstrated in 70% of patients.
5. In most patients an antianginal effect was achieved with doses of 10 mg/24 h.

It may be summarized that transdermal nitroglycerin discs (patches) are an effective and well-tolerated modality of antianginal prophylaxis, at least during daytime, in the hours important for such an effect. As with any other form of drug therapy, there is interindividual variability in response, and the therapeutic regimen should be individually adjusted.

23.3.1. Intermittent Transdermal Nitroglycerin Therapy

Attempts to prevent tolerance to nitroglycerin administered transdermally went along the line of tolerance prevention with long-acting nitrates. This topic is discussed in detail in chapter 7. The most important concept is to avoid constant plasma levels over 24 h. One of the ways to achieve this is to apply the transdermal nitroglycerin patch for a limited period, such as 12 h per day, after which it is removed and a nitrate free interval is allowed until application of a new patch on the next day. A modification of this approach is using a newly developed patch which releases the drug over a limited number of hours per day. With this new patch the early removal of the patch is avoided, thereby maintaining the strong placebo effect of the patch.

The original concept behind the development of transdermal nitroglycerin patches was to maintain constantly high "therapeutic" plasma levels of nitroglycerin. The success of intermittent therapy indicates that this approach might have been inappropriate for some of the patients, who develop tolerance. Modification of this approach by intermittent therapy maintains its major component, therapeutic plasma levels over the active daytime, when angina is more frequent, and a drug-free interval during the night, to avoid tolerance.

Several studies in small series demonstrated the efficacy of intermittent therapy. Schaer et al [348a] reported that exercise tolerance was significantly improved in 11 patients tested 4 and 8 h after application of transdermal nitroglycerin for 14 h per day. The minimal dose used was 20 mg per day, but it was attempted to titrate the patients to the highest tolerance dose. Cowan et al [348b] reported that the antianginal and anti-ischemic effects of continuous therapy were abolished after 1 week in 11 patients with angina pectoris. Intermittent therapy for 12 h per day maintained the efficacy. Luke et al [348c] reported similar findings.

Rudolph et al, who performed many of the studies in the field of tolerance prevention, have recently shown that by increasing plasma concentration of nitroglycerin by repeated transdermal application the marked anti-ischemic effect of this drugs can be maintained nearly unchanged for 12 h [348d]. With a nightly patch-free interval of 12 h tolerance development was avoided and full responsiveness to reapplication on the following day was enabled.

The largest series of patients in whom intermittent therapy had been evaluated was reported by DeMots and Glasser [348e]. In this study 206 patients with chronic stable angina pectoris were evaluated, using serial treadmill testing. After a placebo run-in period the patients were randomized to 4 weeks of double-blind treatment with transdermal nitroglycerin, 5 or 10 mg/day (group A), 15 or 20 mg (group B) or placebo. The patches were applied at 8 a.m. and removed after 12 h.

On the first day of treatment exercise duration increased in both active treatment groups, compared with placebo. After 2 and 4 weeks of treatment exercise duration was again significantly greater for group B than for placebo at 4 and 8 h after application. In group A the initial effect was largely lost. It may be concluded that intermittent therapy with a medium dose of transdermal nitroglycerin results in a sustained effect, without development of tolerance. The difference between the two groups in this study might have been due to the greater use of beta-adrenoreceptor blockers in the low-dose group.

In at least one large-scale trial no difference was found between the efficacy of continuous and intermittent therapy. The Dutch Nitro Study Group [348f] tested the efficacy of transdermal nitroglycerin, 10 mg/24 h, in a randomized, double-blind, placebo-controlled trial in patients with severe angina. After a one week placebo run-in period 127 patients were randomized to placebo, or to intermittent or continuous use of transdermal nitroglycerin. Exercise test was performed prior to randomization and on days 1 and 14 of treatment. Placebo had no effect on exercise duration and time to 1 mm ST depression. Intermittent therapy increased exercise duration from 395 ± 104 to 475 ± 171 seconds on day 1 and to 468 ± 143 seconds on day 14. The values for continuous therapy were 406 ± 115, 468 ± 158 and 451 ± 146 sec, respectively. Changes in time to 1 mm ST-depression paralleled those in exercise of transdermal nitroglycerin increased exercise capacity and prolonged time to onset of ischemia without tolerance during long-term therapy [348f].

Several questions and concerns regarding intermittent patch therapy have been raised:

1. Does the provision of a nitrate-free interval compromise therapy? In a few patients with equal distribution of anginal episodes throughout the day this might be the case. These patients may benefit from combined therapy with other antianginal drugs, given during the nitrate-free interval. However, in most patients the majority of ischemic events occur during the day, and mainly between 8 a.m. and 2 p.m. These patients are well protected by intermittent transdermal nitroglycerin monotherapy.
2. Does the frequent withdrawal of therapy cause rebound phenomena? The Transderm-Nitro Trial Study Group [348e] reported one death and 9 patients with increasing nonexertimal angina during the drug-free interval among the 138 patients assigned to active treatment. These 9 patients completed the study uneventfully.
   An interesting finding of this trial was larger exercise duration in the placebo group than in the active treatment group in the exercise test performed 24 h after patch application, just before the next patch application. It is

possible that patients receiving active therapy felt better with the patch on and were less motivated to exercise without the patch. Alternatively, this finding may be a rebound phenomenon.

3. Can sublingual nitroglycerin be taken during intermittent therapy without interfering with the beneficial effect of the drug-free interval in preventing tolerance? This question has not been addressed in a controlled trial. Retrospective analysis of our experience and of the literature indicates that intermittent therapy maintains its efficacy even if sublingual nitroglycerin is taken to control acute anginal episodes during the drug-free interval.

Administration of high doses of long-acting nitrates is one of the most important ways of inappropriate use, resulting in tolerance. This is the case also with transdermal nitroglycerin. Intermittent therapy can prevent the development of tolerance when low to medium doses are used but fails to do so for high doses. Nabel et al [348g] studied by ambulatory electrocardiographic monitoring, 14 patients with chronic stable angina pectoris who were titrated to a maximally tolerated dose of 30–60 mg/day (mean dose 52 ± 5 mg). Two crossover double-blind placebo-controlled phases were randomly used: continuous therapy and intermittent therapy for 12 h per day. On the first day of treatment both regimens significantly reduced the frequency of ischemic episodes, but they returned to placebo level on the second day. Thus with very high doses of transdermal nitroglycerin the anti-ischemic effect is lost due to rapid development of tolerance. Intermittent therapy does not prevent this loss.

A new patch with a phased-release mechanism is now being evaluated. It is expected to solve the problem of tolerance due to constantly high plasma levels of nitroglycerin.

In summary, intermittent therapy with drug-free interval of 10 to 12 h can prevent tolerance to transdermal nitroglycerin in patients who develop such tolerance during continuous therapy. As with other nitrate formulations, it is advisable to use low to medium doses. The tolerability of this regimen is generally good.

23.4. Unstable Angina

Flaherty [349] has recently studied the crossover from intravenous to transdermal nitroglycerin in 17 patients with unstable angina pectoris. Intravenous nitroglycerin was initially titrated to reduce mean arterial pressure by 10% and further titrated to reduce episodes of chest pain and electrocardiographic changes. Thereafter the infusion rate was reduced by 50% and 5 or 10 mg/24 h transdermal nitroglycerin were applied. Additional doses were given every 30 min, and the intravenous nitroglycerin infusion rate was reduced and finally discontinued.

Fifteen of the 17 patients had relief of pain with intravenous nitroglycerin. Ten of these patients, with coronary artery disease demonstrated by catheterization, remained asymptomatic during the first 24 h after crossover. Seven of them remained asymptomatic until discharge.

These findings indicate that transdermal nitroglycerin in individually titrated doses can sustain the beneficial anti-ischemic effect of intravenous nitroglycerin in patients with unstable angina.

23.5. Silent Ischemia

As silent ischemia is asymptomatic, treatment should be based on a constant therapeutic level of the drugs used in the plasma. If nitrate therapy is considered, one of the best alternatives are nitroglycerin discs. Shell et al [350] studied eight patients with chronic stable angina pectoris and silent ischemia evident by ST-segment depression or elevation on ambulatory 24-h electrocardiogram. The patients were treated with beta-blockers to inhibit demand angina. Transdermal nitroglycerin was started at a dose of 5 mg/24 h, and the dose was increased at increments of 5 mg. At a mean dose of 10.4 mg/24 h all eight patients had a decrease in the number of symptomatic events. The total number of silent and symptomatic ischemic events fell from 5.3/24 h to 0.8/24 h. (About 80% of the events were silent.) The duration of ischemic events decreased from 95.8 min/24 h to 17 min/24 h. The magnitude of ischemia, as estimated from the integral of ST-segment deviations also decreased.

Thus, transdermal nitroglycerin in the presence of beta-blockade, can reduce the number, duration and magnitude of silent as well as symptomatic ischemic events.

It is important to evaluate the effect of transdermal nitroglycerin also in the absence of beta-blockade. My colleagues and I have started a study of transdermal nitroglycerin, monotherapy, in elderly patients with silent ischemia or atypical symptomatic presentations of ischemic events. Preliminary findings are promising, but we must still study a larger number of patients. Deponit® was found effective in abolishing silent ischemia in patients with angina pectoris who had silent ischemia despite treatment with beta-blockers and calcium-antagonists.

23.6. Congestive Heart Failure

Transdermal nitroglycerin patches are suitable for ambulatory treatment of patients with congestive heart failure. Hemodynamic and clinical improvement has been observed in short-term and chronic studies. In most reported studies of transdermal nitroglycerin in heart failure the hemodynamic evaluation was limited to 24 h.

Olivari et al [351] studied the hemodynamic effects of transdermal nitroglycerin, 15–20 mg/24 h, in nine patients with heart failure. The drug produced a sustained hemodynamic response which began 30 min after application and fully persisted for at least 6 h. A significant decrease in right and left ventricular filling pressures was associated with an increase in stroke index and a significant decrease in systemic and pulmonary vascular resistances. There was no change in heart rate and systemic arterial pressure. After 24 h pressures had partially returned to control levels, but mean pulmonary artery pressure was still significantly lower than in the control period.

In contrast, Rajfer et al [329], also studying nine patients with heart failure, found that the hemodynamic response, which was observed at 0.5–1.0 h and peaked at 6 h, persisted for 24 h. However, they stated that the majority of patients required relatively large doses of transdermal nitroglycerin to produce a satisfactory improvement in cardiac performance that was sustained for 24 h. In none of these studies were the hemodynamic parameters continuously measured over more than 1 day at the initiation of treatment.

Sharpe and Coxon [352] reported that reduction in pulmonary capillary wedge pressure, produced by 5 and 10 mg/24 h of transdermal nitroglycerin persisted for 24 h.

These investigators treated ten patients who initially responded to transdermal nitroglycerin, 5 mg/24 h for 3 months. The reduction in systemic vascular resistance was maintained throughout this period, but the reduction in pulmonary capillary wedge pressure and to some extent also the other hemodynamic changes, were attenuated.

This study is the only one evaluating the hemodynamic response also in the 2nd day after initiation of treatment. It demonstrated that the reductions in pulmonary capillary wedge pressure and pulmonary arterial pressure were maintained on the 2nd day of treatment, but the changes in cardiac index, stroke volume index and systemic vascular resistance were attenuated, indicating rapid tolerance to the arteriodilating effect of nitroglycerin. However, the doses used in this study (5 and 10 mg/24 h) were lower than those recommended today for severe heart failure. As it is important to evaluate the hemodynamic response to transdermal nitroglycerin for more than 24 h, and as the 2nd day of treatment is particularly important, my colleagues and I studied the hemodynamic response to transdermal nitroglycerin patches by continuous hemodynamic monitoring over 72 h. Eight patients with chronic congestive heart failure and functional class III were studied. Seven of them were men, and one was a woman, of mean age 66 $\pm$ 10 years. The cause of heart failure was ischemic heart disease with anterior wall myocardial infarction in all patients. Three patients had also inferior wall infarction. Six of them were in sinus rhythm, and two had atrial fibrillation. All were treated with digoxin and diuretics until 24 h prior to the study. All other drugs (nitrates, captopril and nifedipine) were discontinued at least 4 days prior to the study.

The patients were hospitalized in the coronary care unit, a Swan-Ganz catheter was introduced, and a placebo patch was applied. Cardiac pressure was measured, cardiac output was determined by the thermodilution technique, and systemic and pulmonary vascular resistances were measured at 1 and 2 h after application of the placebo patch. Transdermal nitroglycerin patches, 10 mg/24 h, were then applied, and the dose was titrated upward until the patient developed headache, or until mean systemic arterial pressure was lowered by 10 mmHg. Systolic blood pressure was not lowered beyond 90 mmHg. Hemodynamic measurements were performed every hour in the first 4 h and every 4 h thereafter for 72 h. The patches were replaced by the same dose every morning. The paired t-test was used for evaluation of the statistical significance of the results. The dose used was 30 mg/24 h in six patients and 20 mg/24 h in two. On the 1st day transdermal nitroglycerin produced hemodynamic improvement in all patients. The effect was evident at 2 h and reached a peak at 6 h. Pulmonary arterial pressure, pulmonary capillary wedge pressure, right atrial pressure and systemic and pulmonary vascular resistances decreased, and cardiac output increased. Heart rate and systemic arterial pressure were not significantly altered (Table 1). The effect was maintained, without significant changes, throughout 24 h.

In the whole group the hemodynamic response was partially but significantly attenuated during the 2nd day and it remained at this level during the 3rd day. However, the hemodynamic changes, compared with pretreatment measurements, were still significant at the 2nd and 3rd days, except for the decrease in right atrial pressure which was totally abolished. There was no significant difference between the 2nd and 3rd days.

When the patients were analyzed separately, it was shown that in four of them there was no tolerance, the initial hemodynamic response being sustained almost completely over three days. Three patients developed tolerance to all hemodynamic effects on the 2nd day, and the response remained attenuated during the

Table 1. Hemodynamic parameters in eight patients before treatment, on the 1st day and on the 3rd day (no significant difference was found between the 2nd and 3rd day)

	Baseline	**Day 1**	$p<$	**Day 3**	$p<$
BPS	111.8 ± 11.7	103.7 ± 9.9	N.S.	110 ± 8.2	N.S.
BPD	66.2 ± 6.9	61.2 ± 5.9	N.S.	66.8 ± 7.0	N.S.
BPM	81.4 ± 7.7	75.4 ± 6.8	0.05	81.2 ± 6.7	N.S.
PAS	56.0 ± 11.9	46.1 ± 8.5	0.01	48.8 ± 7.4	0.02
PAD	23.0 ± 3.2	14.8 ± 3.3	0.01	18.6 ± 2.8	0.05
PAM	34.0 ± 5.2	25.3 ± 4.8	0.02	28.7 ± 3.4	0.05
PCWP	21.5 ± 1.9	14 ± 2.9	0.01	17.5 ± 2.3	0.05
CO	3.5 ± 0.5	4.0 ± 0.4	0.05	3.7 ± 0.4	0.01
SVR	1781 ± 335.3	1449 ± 215	N.S.	1648 ± 215	0.02
PVR	303.5 ± 141.8	229 ± 62	0.05	238 ± 44	0.05
RA	5.5 ± 2.0	3.8 ± 1.0	0.05	4.6 ± 1.1	N.S.

Table 2. Hemodynamic changes in patients with sustained response

	Baseline	**Day 3**	$p<$
BPS	115 ± 11.1	107.5 ± 11.4	0.005
BPD	67.5 ± 4.3	65.0 ± 3.5	0.01
BPM	84.5 ± 6.6	79.1 ± 6.0	0.005
PAS	58.7 ± 13.9	51.2 ± 9.6	0.02
PAD	23.0 ± 3.4	17.5 ± 2.8	0.005
PAM	34.9 ± 6.7	28.7 ± 5.1	0.005
PCWP	21.2 ± 1.9	16.0 ± 2.1	0.005
CO	3.2 ± 0.5	3.6 ± 0.4	0.05
SVR	1924.1 ± 152.0	1648.5 ± 167.5	0.05
PVR	350.1 ± 157.3	278.8 ± 66.7	0.02
RA	5.2 ± 2.1	4.2 ± 0.8	0.01

Table 3. Hemodynamic changes in patients with attenuated response

	Baseline	Day 3	$p<$
BPS	106.2 ± 9.6	106.6 ± 6.2	N.S.
BPD	62.5 ± 4.3	63.3 ± 4.7	N.S.
BPM	77.1 ± 4.7	77.7 ± 2.0	N.S.
PAS	50.0 ± 9.2	52.0 ± 2.4	N.S.
PAD	21.5 ± 4.3	18.0 ± 1.6	N.S.
PAM	31.0 ± 4.4	29.3 ± 0.2	N.S.
PCWP	20.2 ± 3.5	17.0 ± 0.8	N.S.
CO	3.8 ± 0.3	4.0 ± 0.1	N.S.
SVR	1496.8 ± 179.6	1448.0 ± 17.7	N.S.
PVR	230.9 ± 97.6	243.5 ± 16.3	N.S.
RA	5.5 ± 2.06	4.3 ± 0.4	N.S.

BPS, Systolic blood pressure (mmHg); *BPD,* Diastolic blood pressure (mmHg); *BPM,* Mean blood pressure (mmHg); *PAS,* Pulmonary artery systolic pressure (mmHg); *PAD,* Pulmonary artery diastolic pressure (mmHg); *PAM,* Pulmonary artery mean pressure (mmHg); *PCWP,* Pulmonary capillary wedge pressure (mmHg); *CO,* Cardiac output (L/min); *SVR,* Systemic vascular resistance (dyn sec cm^{-5}); *PVR,* Pulmonary vascular resistance (dyn sec cm^{-5} m^{-2}); *RA,* Right atrial pressure (mmHg)

3rd day. One patient developed tolerance on the 1st day, and the response remained attenuated during the 3rd day. One patient developed tolerance on the 1st day, and the response remained attenuated during the 2nd and 3rd days (Tables 2 and 3). One patient developed slight systemic hypotension at the initiation of treatment. He responded favorably to a reduction in dose. No other serious adverse effects were observed.

Our results indicate that transdermal nitroglycerin at doses of 20–30 mg/24 h produces hemodynamic improvement in patients with congestive heart failure. This improvement was maintained in 50% of the patients throughout 3 days of continuous hemodynamic monitoring. In the remaining patients there was tolerance to the hemodynamic effect, which became evident usually during the 2nd day of treatment but in one patient even on the 1st day. The responders could not be differentiated from the nonresponders by any clinical or hemodynamic parameter at the initiation of study.

These findings show that monitoring for 24 h is not sufficient to detect early tolerance in most cases, and 48 h will be preferable. On the other hand, monitoring for 72 h is probably not required.

Thus our study was the first to evaluate, by continuous hemodynamic monitoring for 72 h, the currently recommended doses of transdermal nitrogylcerin in heart failure.

Our results and those of several of the studies mentioned above contradict those of a recent study by Armstrong [353] which revealed that transdermal nitroglycerin, at doses of 10–30 mg/24 h, has minimal hemodynamic effects in patients with advanced heart failure. This difference may be attributed in part to the different patients studied by Armstrong and by us. The incidence of sustained response and tolerance in our series is, however, similar to that observed by Elkayam et al [354] with intravenous nitroglycerin.

Recently Cowan et al [355] reported that the antianginal effect of transdermal nitroglycerin was abolished during continuous therapy but was fully maintained with intermittent therapy with a nitrate-free interval of 12 h. This might be valid also for congestive heart failure. However, in patients with heart failure it might be preferable not to leave the patient without vasodilators for 12 h. Therefore we studied the efficacy and possibility of early tolerance to combined treatment with transdermal nitroglycerin during the night and ISDN during daytime by continuous hemodynamic monitoring in eight patients with severe chronic congestive heart failure.

All drugs excepting digitalis and diuretics were discontinued, and baseline hemodynamic parameters were determined on the placebo patch. The patients then received, at 8 p.m. 10 mg transdermal nitroglycerin. At 8 a.m. the patch was removed, and the patients received isosorbide dinitrate, 40 mg orally, at 8 a.m. and at 3 p.m. Hemodynamic measurements were performed every 3 h, for 3 days. The drugs produced marked hemodynamic improvement which was sustained throughout the period of study. The maximal effect became evident within 2 h from application of the patch. The pulmonary capillary wedge pressure decreased from 21.5 ± 1.9 to 14 ± 2.9 mmHg, cardiac output increased from 3.5 ± 0.5 to 4.0 ± 0.4 L/min, and right atrial and pulmonary arterial pressures and systemic and pulmonary vascular resistances decreased. All these changes were statistically significant. They were maintained throughout the study period, despite discontinuation of diuretics in four of the patients and a decrease in dose in the other four. Symptomatic improvement was observed in all patients. They remained clinically well without diuretics or with the lower dose of diuretics, as determined during the study, throughout 2 months of follow-up on the combined nitrate therapy.

These findings indicate that combined therapy with transdermal nitroglycerin and isosorbide dinitrate produce sustained hemodynamic and clinical improvement. This regimen may be an alternative to diuretics in some patients.

It may be questioned whether the hemodynamic effect resulted mainly from ISDN. This is not likely, however, because the afternoon dose of ISDN was given at 3 p.m. and could not produce a hemodynamic effect for 16 h until the administration of the morning dose at 8 a.m. [356]. The hemodynamic effect showed no significant attenuation during the night. Therefore it might be attributed mainly to the transdermal nitroglycerin.

An important by-product of this study was the finding that diuretics might be discontinued or reduced in dose in most patients with heart failure who receive adequate vasodilator regimens. In view of the more advantageous hemodynamic profile of vasodilators in heart failure (particularly in respect to the increase in cardiac output) [357] and the well-known complications of diuretics, mainly in elderly patients, nitrates may be considered an alternative to diuretics in first-line treatment of heart failure.

In summary, we have shown by direct and continuous hemodynamic monitoring that combined treatment with transdermal nitroglycerin during the night and ISDN orally during daytime produces sustained hemodynamic improvement in congestive heart failure.

The clinician treating ambulatory patients with congestive heart failure now has two alternatives for simple first-line vasodilator therapeutic regimen:

1. Transdermal nitroglycerin patch.
2. Once-daily administration of sustained release ISDN or IS-5-MN.

Sharpe et al [357a] compared the hemodynamic effects of intermittent and continuous treatment with transdermal nitroglycerin, 10 mg/24 h, in 10 patients with chronic congestive heart failure. In 8 of them the initial patch application resulted, within 2 h, in more than 20% reduction in pulmonary capillary wedge pressure, from 22 ± 5 to 16 ± 6 mmHg. Cardiac index increased in these 8 patients from 2.1 ± 0.5 to 2.4 ± 0.6 L/min · m^2. The two patients who did not show such response had the largest left ventricular volumes in this group. The responders were randomized to intermittent (8 h/day drug-free interval) or continuous treatment for one month, and then crossed over to the alternate treatment for another month. Cardiac catheterization was repeated at the end of each month. At the end of one month of intermittent therapy the hemodynamic response after reapplication was similar to the initial response. After one month of continuous therapy the hemodynamic values 24 h after patch application were similar to pretreatment values. Removal and reapplication of the patches resulted in no response. It was concluded that the vasodilator effect of transdermal nitroglycerin in congestive heart failure is maintained during intermittent treatment, whereas tolerance develops during continuous treatment.

Further studies are required to determine the superiority of either method.

Transdermal nitroglycerin discs have no special advantage for patients with acute myocardial infarction, and they were not designed for use in this condition. Nevertheless, Nelson et al [358] studied the hemodynamic effects of transdermal nitroglycerin, 50 mg, in ten patients with pulmonary edema (left ventricular filling pressure over 20 mmHg) not associated with hypotension, which developed 6–18 hours after acute myocardial infarction. No effects were noted until 60 min after administration of the drug, when slight falls in systolic and diastolic blood pressures were observed. Slight falls in left ventricular filling pressure, by 3 mmHg, were evident at 90 min and persisted until 3 h after administration. Heart rate and cardiac output were not altered.

It was concluded that the effects of transdermal nitroglycerin, even at the high dose of 50 mg, were too slow in onset, too small in magnitude and of too short duration to be of any clinical significance.

Isometric exercise may cause hemodynamic deterioration in patients with severe congestive heart failure due to an increase in both left ventricular preload and afterload. Although nitrates have been found effective in the treatment of hemodynamic impairment during dynamic exercise, their effect during isometric exercise has not been investigated. Recently Roth et al (359) studied the acute effect of 120 mg transdermal nitroglycerin on hemodynamic changes induced by isometric exercise in 18 patients with severe congestive heart failure. The drug produced a significant hemodynamic improvement during isometric exercise, as evidenced by decreases in left and right ventricular preload and afterload and increases in cardiac output and stroke work index. Heart rate was not significantly altered. Mean blood pressure increased from 94 mmHg at rest to 101 mmHg during isometric exercise without treatment but was not altered during exercise on transdermal nitroglycerin. Right atrial pressure increased from 10 to 14 mmHg without treatment but was not altered on nitroglycerin. Pulmonary capillary wedge pressure increased from 27 mmHg at rest to 31 mmHg during pretreatment isometric exercise on nitroglycerin. Cardiac index slightly decreased during pretreatment exercise but returned to resting values during isometric exercise and decreased to 22 mmHg during exercise on nitroglycerin. It was concluded that the direct vasodilatory effect of nitroglycerin counteracts the hemodynamic deterioration produced by isometric exercise in patients with congestive heart failure.

The use of transdermal discs has again raised the possibility of development of tolerance to the hemodynamic effect of nitrates. It was suggested that the prolonged constant release of nitroglycerin and constant plasma levels may accelerate tolerance, perhaps by regulatory effects on the receptors for nitrates. It has been shown that after short periods (hours) of transdermal nitroglycerin administration counterregulatory mechanisms are activated [360]. These mechanisms might be those related to attenuation of the effect of vasodilators in general (but acting more rapidly) or mechanisms specifically related to the receptors for nitroglycerin.

Cohn stated that transdermal preparations should be used in "exceedingly large doses" to achieve an effect comparable to that of oral dosing and that "the chronic response to these preparations has not yet been carefully assessed" [361]. In my opinion the problem has been approached from a relatively mistaken direction. It is suggested that if rapid onset of tolerance to constant administration of nitrates is a real phenomenon, then the concept of transdermal nitroglycerin disc is wrong [330]. It has been attempted to show that an interval without nitrates is required to maintain the efficacy of these agents [362, 114]. The simple possibility of detachment of the disc before going to bed in the evening to produce a 5-h to 10-h nitrate-free interval during the night has been usually ignored. Only recently have studies showing the efficacy of a nitrate-free interval in preventing tolerance been reported.

23.7. Potential Mechanisms for Tolerance

Several mechanisms have been suggested to account for the cases of partial or complete attenuation of the effects of nitrates in some patients.

When it develops, tolerance appears to be related to the dose used, the extent of changes in plasma concentrations of the drug, and the half-life of the specific nitrate used. Some of these are nonspecific mechanisms, related to activation of compensatory mechanism by nitrates as well as by other vasodilators. Even if these mechanisms result in attenuation of the effect they are less problematic with nitrates than with other vasodilators, such as some of the alpha-adrenoreceptor blockers.

Specific mechanisms for nitrates have also been suggested. Perhaps the most well-known is the decrease in reduced sulfhydryl groups. These groups, located in vascular smooth muscle (and in other sites) can contribute glutathione for activation of organic nitrates, required for their vasodilatory effect (by cGMP) [363, 364].

Recently it has been suggested that plasma volume expansion may explain tolerance to nitrates. This theory has been specifically studied for transdermal nitroglycerin [365]. It is well-known that intravenous administration of nitroglycerin over 24 h produces significant reduction in hematocrit, implying plasma volume expansion. The expansion results from the continuous vasodilation, which leads to perturbation of intracapillary starling forces. Barclay et al [365] compared the effects of transdermal nitroglycerin and placebo over 24 h, using a double-blind, crossover protocol in six healthy volunteers. Nitroglycerin reduced the hematocrit by 5% at 3 h, and this effect was maintained for 24 h. The colloid osmotic pressure was similarly affected. The diameter of the superior mesenteric artery and of the hepatic artery was increased by 15%. The left ventricular end-diastolic dimension was reduced maximally by 7% at 7 h and returned to baseline at 24 h. Blood pressure fell by 8% at 3 h, returned to baseline at 7 h and was slightly elevated at 24 h.

It was concluded that pharmacologic tolerance did not occur, as both the superior mesenteric and hepatic arteries remained dilated. However, the hemodynamic effect was attenuated, in association with plasma volume expansion. This mechanism may contribute to the development of tolerance to transdermal nitroglycerin.

23.8. Pharmacokinetics of Nitroglycerin Discs

Nitroglycerin easily crosses the skin and is absorbed into the systemic circulation, bypassing the hepatic circulation. Therefore its delivery has near zero-order kinetics, similar to an intravenous infusion at a constant rate. The rate of nitroglycerin delivery is determined by the size of the disc applied to the skin. Delivery begins immediately after application of the disc, steady-state plasma concentration being achieved at about 2 h [338]. Conventional discs achieve relatively low plasma concentrations. Delivery rates of commercially available discs are about 0.5–0.625 mg cm^{-1} 24 h^{-1}.

Abrams [330] believes that one of two major problems of transdermal nitroglycerin discs are the relatively low plasma levels achieved by conventional doses. Müller et al [338] reported that a dose of 10 mg daily achieved plasma levels of less than 0.5 ng/ml, and one of 20 mg daily achieved levels of less than 0.7 ng/ml.

Recently Zwicke et al [366] have shown that the rate of nitroglycerin absorption is lower and the total quantity absorbed from transdermal discs less in Blacks than in Caucasians. This observation should be considered when determining dosage of transdermal nitroglycerin for clinical application.

Barkve et al [367] studied the pharmacokinetics of nitroglycerin in the form of transdermal discs, 10 mg, in 12 healthy volunteers during exercise and under conditions of high ambient temperature. During treatment under control conditions mean nitroglycerin plasma concentration was 1.0 to 1.5 nmol/L, after 20 min of exercise it was 3.1 ± 1.7 nmol/L, and after exposure to 90° C for

20 min the plasma concentration was 7.3 ± 1.7 nmol/L. Compared with control levels these differences were statistically significant. These findings suggest increased absorption of transdermal nitroglycerin during work. This may be beneficial for the exercising patient but may be undesired in a hot climate or in hot surroundings.

23.9. Transdermal Nitroglycerin Discs - What Do We Know Now?

At present nitroglycerin discs are widely used. Their use is based on both proven clinical efficacy and many clinical studies. These studies have frequently, but not always, revealed efficacy in angina pectoris and congestive heart failure. In a recent review Charash and Scheidt [345] stated that eight of ten acute studies demonstrated favorable results at 2–12 h and five of eight studies at 24 h. It is more difficult to interpret the results of chronic studies, but it is important to note that only two of nine studies have shown completely negative results. These are the studies of Parker and Fung [115] and Sullivan and coworkers [368]. Moreover, most long-term studies have shown decreases in anginal frequency and sublingual nitroglycerin consumption.

These data, together with the excellent acceptability of transdermal nitroglycerin discs by patients and physicians, indicate that the discs are an important and effective therapeutic modality.

23.10. ISDN Ointment

ISDN ointment is also available for clinical use, although at present it is seldom used. It has been found very effective in treatment of angina pectoris. It was developed to avoid first-pass hepatic degradation of the drug. Repeated applications of this ointment at intervals of 4–5 hours produce gradual elevation of plasma concentration, until a steady state is achieved at the end of the 2nd day. The drug is absorbed from the skin at a relatively constant rate. However, a marked interpatient variability in the rate of absorption has been observed.

The hemodynamic effects of ISDN ointment resemble those of other forms of the drug. Imhof et al [369] reported that the hemodynamic effects of ISDN ointment were of a lesser magnitude than those of nitroglycerin ointment. Side effects were also less pronounced.

The drug may be detected in the plasma for 24 h after application. This probably results from continued absorption from a cutaneous reservoir [369]. Parker et al [114] reported that ISDN ointment increased exercise duration in 11 patients with chronic stable angina pectoris from 349 ± 107 to 457 ± 151 sec. The effect was evident 2 h after administration of the drug and persisted for 8 h. No effect was observed at 24 h. Distante et al [192] reported that chronic administration of ISDN ointment produced sustained antianginal effect despite the development of hemodynamic tolerance.

Handler and Sullivan [370] compared ISDN cream, 50 mg twice daily, to sustained-release ISDN tablets, 20 mg twice daily, in 28 patients with coronary artery disease and stable angina pectoris. The patients received for 1 week only sublingual nitroglycerin, and they were then randomly allocated to either cream or tablets for 2 weeks, followed by the alternate preparation for another 2 weeks. Both preparations increased the mean treadmill exercise time to onset of angina pectoris and to termination of exercise to the same extent compared with pre-treatment values. There were no differences between the two preparations regarding frequency of anginal episodes, nitroglycerin consumption, heart rate and ST-segment depression at onset of angina and the double product and ST-segment depression at maximal exercise. Equal numbers of patients preferred to continue treatment with either preparation.

In conclusion, ISDN ointment is as effective as sustained-release ISDN tablets in controlling anginal episodes and suppressing myocardial ischemia in patients with coronary artery disease [370].

23.11. Transdermal ISDN Spray

A transdermal ISDN spray which keeps plasma ISDN levels in the therapeutic range for 8 to 12 h, is now available. The dose used is 60 mg once or twice daily. The first dose of this spray was shown to elicit the expected hemodynamic and electrocardiographic changes in patients with angina pectoris. However, tolerance developed within 2 weeks even in patients who received only one daily dose [370a].

23.12. Cutaneous Properties Relevant for Transdermal Therapy

Until recently the skin has been considered an impermeable barrier. Dermatologic research was preoccupied with attempts to locate the presumed layer which is responsible for impermeability, thereby delaying investigation on transdermal delivery of drugs. Only in the past two decades have such studies been extensively performed. The features of the skin as a barrier and as a potential portal of entry for drugs depend almost exclusively on the stratum corneum, which comprises most of the mass of the epidermis. This layer prevents the evaporation of water through the skin. Many lipophil compounds can cross the stratum corneum and reach the systemic circulation. However, the rate of penetration is slow, limiting the rapidity of drug delivery through the skin. Another important feature of the stratum corneum is its reservoir capacity. Lipophil substances may be absorbed by this layer, achieving relatively high concentrations. This reservoir slowly releases the substances to the systemic circulation.

High molecular weight substances and water-soluble substances penetrate the skin poorly. Lipid-soluble substances of low molecular weight are most likely to penetrate the skin.

Skin permeability is determined by thickness of the stratum corneum. Suntanning thickens this layer. Thickness of the stratum corneum is not altered with age. In most clinically available transdermal nitroglycerin devices the properties of the skin, and particularly the stratum corneum, are the main regulators of flux. In modern devices which are now under development the rate of flux is controlled by the device itself.

24. Buccal Nitroglycerin

An interesting attempt to use nitroglycerin for long periods and avoid firstpass hepatic elimination was made by the development of buccal nitroglycerin tablets. Only a few clinical studies have been reported with this method, and it is seldom used in clinical practice. Nevertheless it deserves description because it emphasizes some of the problems of nitroglycerin therapy for long durations and suggests solutions for them. Although transdermal administration of nitroglycerin is the method most widely used for long-term nitroglycerin therapy, buccal administration may be useful in some patients with cutaneous problems.

Buccal nitroglycerin tablets are composed of an inert polymer matrix containing the drug. The tablet is placed in the buccal cavity between the upper lip and the teeth. It is adhered to the mucosa by formation of a coating by the tablet's surface. The polymer matrix allows continuous diffusion of nitroglycerin from the tablet. Onset of action is within 2–10 min from administration, and the delivery of nitroglycerin continues as long as the tablet remains in place. The plasma levels of nitroglycerin are directly related to the rate of dissolution of the tablet. When the tablet is removed, nitroglycerin rapidly disappears from the plasma. The hemodynamic effect of nitroglycerin persists for about 30 min after removal of the tablet [371–374]. The tablet dissolves within 1–6 hours, depending on various local factors in the mouth. Thus, buccal nitroglycerin with its rapid onset and long duration of action combines the beneficial properties of sublingual nitroglycerin and long-acting organic nitrates.

Buccal nitroglycerin was found effective in several clinical studies of patients with angina pectoris. Relief of pain was achieved within 2 min after administration, almost as rapidly as conventional sublingual nitroglycerin tablets. Exercise performance was increased by the drug within 30–60 min after administration. This effect was equal to that produced by sublingual nitroglycerin 3 min after administration [375]. Patients with angina pectoris, in whom the tablet remained adherent to the mucosa for 5 h or more, showed an improvement in exercise performance for 5 h after administration.

The duration of antianginal effect of buccal nitroglycerin is a function of the dose. It was reported that a 1-mg dose produced an antianginal effect for 1 h and a 2.5-mg dose for 3 h.

Buccal nitroglycerin can be used also in congestive heart failure. In this condition the drug produced symptomatic relief and prolonged exercise duration throughout treatment for 1 month [376].

Adverse effects of buccal nitroglycerin result mainly from vasodilation and are similar to those of other nitrates. A special risk, however, results from the fact that the patient holds in his mouth a tablet containing a high dose of nitroglycerin - about 10 times higher than that of a conventional sublingual nitroglycerin tablet. If the patient, by mistake, chews the tablet and swallows it, the drug is absorbed in the intestine and undergoes first-pass hepatic metabolism. However, if the patient chews the tablet, but does not swallow it, a high dose of nitroglycerin can be rapidly absorbed by the mucosa and reach the systemic circulation without first-pass hepatic metabolism. This high dose may have deleterious effects.

The drug does not cause any local effects at the site of adherence, but about half the patients have difficulties in eating and/or drinking. The main adverse effects are headache and orthostatic symptoms. In one study headache was found in 27% of patients and orthostatic symptoms in 8% [377].

D. Pharmacology

25. Clinical Pharmacology

The pharmacokinetics of nitrates differ markedly and should be discussed separately for each agent. In this chapter the pharmacokinetic profiles of the three most widely used nitrates, namely, nitroglycerin, ISDN, and IS-5-MN are discussed.

25.1. Nitroglycerin

Nitroglycerin is administered by more routes than any other cardioactive drug. It may be given sublingually, orally, intravenously, transdermally, buccally, rectally, by inhalation or directly into the coronary arteries. Although the drug has been used for about 100 years, its pharmacokinetic profile has been extensively studied only in the past 15 years, since the development of accurate methods for its identification and quantification in the plasma. These studies have been performed in a suprisingly small number of patients.

The most rapid conventional method for administration of nitroglycerin is by sublingual tablets. Several studies in the 1970s confirmed that nitroglycerin can be detected in the plasma shortly after sublingual administration [378, 379, 251, 380]. Armstrong et al [251] reported that nitroglycerin was detected in the plasma 30 sec after sublingual administration. It reached a peak after 5 min, fell to 50% of maximal concentration at 7.5 min and was hardly detectable within 20 min from oral administration. Peak plasma concentrations obtained with sublingual nitroglycerin, 0.6 mg, were 2–3 ng/ml.

Wei and Reid [381] studied the time course of plasma concentrations of nitroglycerin in young healthy volunteers. Within 2 min after sublingual administration of a 0.6 mg tablet plasma concentrations of 1.1 ± 0.3 ng/ml were measured. Plasma concentration reached a peak of 1.4 ± 0.6 ng/ml within 5 min.

This study also provided the best correlation between plasma levels and the effect of nitroglycerin. The plasma nitroglycerin levels correlated directly with both the maximal effect and the time course of changes in cardiovascular variables that reflect the known responses to this drug. These include heart rate, systolic arterial pressure, left ventricular diastolic and systolic dimensions, left atrial dimensions and velocity of circumferential fiber shortening. The correlation of plasma levels with changes in heart rate was better than with systolic blood pressure. No correlation was found with diastolic blood pressure [381].

In patients with acute myocardial infarction the correlation between plasma concentrations of nitroglycerin and changes in cardiovascular variables was not as strong as in healthy volunteers because of interference of various hemodynamic and neurogenic mechanisms [378].

Zimrin et al [382] studied the correlation between infusion rates, plasma concentrations and hemodynamic effects of intravenously administered nitroglycerin, titrated to maximal tolerated dose in eight patients with chronic stable angina pectoris. Most patients showed strong correlations between infusion rate and blood pressure during titration, but correlations between blood pressure and concentration, heart rate and concentration, and heart rate and infusion rate were weaker. Individual regression lines varied widely. It was concluded that although hemodynamic response to a given infusion rate of nitroglycerin and concentration varies widely among individuals, the relationship of infusion rate, concentration and resting hemodynamic effects of a maximal tolerated dose of intravenous nitroglycerin is constant over 24 h.

Nitroglycerin has no predictable effect after oral administration, and it usually has no effect at all. This is because orally administered nitroglycerin undergoes, similarly to other nitrates, extensive first-pass hepatic metabolism. Attempts have been made to overcome this problem by the use of high doses in oral, slow-release formulations. These formulations are rarely used. Sublingually administered nitroglycerin does not undergo first-pass hepatic metabolism. This is the case also with buccal nitroglycerin, which is discussed in a separate chapter.

There are two methods to achieve constantly high plasma levels of nitroglycerin. Intravenous administration of nitroglycerin is effective within a few minutes from initiation of the infusion and achieves plasma concentrations ranging between 1 and 65 ng/ml [384]. The only practical method to achieve constant plasma levels of nitroglycerin in ambulatory patients is by transdermal administration. Transdermal nitroglycerin discs are the most widely used formulation for transdermal administration, although nitroglycerin ointment is also available. Transdermal devices are discussed in a separate chapter. Systemic bioavailability of transdermal nitroglycerin is about 75%.

The pharmacokinetic profile of nitroglycerin oral spray resembles that of sublingual nitroglycerin. This is discussed in the chapter on nitrate oral spray.

Nitroglycerin has a large volume of distribution. It is eliminated by denitration, mainly in the liver, at a very rapid rate. Total body clearance is about 25–30 L/min, and the elimination half-life is about 4–5 min. The fact that total body clearance is greater than hepatic blood flow indicates that nitroglycerin is also eliminated in other sites. Nitroglycerin is eliminated by glutathione nitrate reductase which was found in the liver and erythrocytes. This denitration results in formation of glyceryl dinitrate and mononitrate which undergo further denitration to glyceral or are excreted by the kidneys as glucoronides [383]. The glyceryl dinitrates have weak vasodilator activity.

The pharmacokinetics of nitroglycerin should be studied specifically in congestive heart failure. Armstrong et al [171] evaluated the pharmacokinetic profile of intravenously administered nitroglycerin in 16 patients with congestive heart failure. A strong correlation was observed between the infusion rate and plasma concentrations. Hemodynamic improvement, evident as a 25% fall in pulmonary capillary wedge pressure, was achieved in patients with plasma concentrations of 1.2–11.1 ng/ml. The minimum effective plasma nitroglycerin concentration was 1.2 ng/ml at an infusion rate of 15 μg/min. Some of the patients did not respond, despite plasma levels higher than 11 ng/ml. These patients had higher right atrial pressures than responders (19 ± 4 versus 10 ± 4 mmHg). Some patients failed to reach the hemodynamic end-point even with high infusion rates of 220–440 μg/min and plasma levels of 42.2–481.3 ng/ml. Systemic congestion may be one of the causes of this refractoriness.

25.2. ISDN

25.2.1. Absorption

ISDN can be given sublingually, orally, transdermally, intravenously or directly into the coronary arteries. It is rapidly absorbed after sublingual or oral mucosal administration. Maximal plasma levels are achieved within 2 min from administration of ISDN oral spray and 2–5 min from administration of sublingual tablets. It is also rapidly absorbed from the gastrointestinal tract after oral administration. The effect is evident within 15–20 min and usually reaches a peak in 1 h. Absorption is almost complete and is not affected by intake of food. In a study by Thadani et al [13] peak plasma concentrations of ISDN were observed within 30–120 min from oral administration. These investigators reported pharmacokinetic differences between acute and chronic treatment: not only did the peak plasma concentrations differ but also the time to peak concentrations. In chronic therapy peak levels of 9.0, 20.4, 67.9 and 85.0 ng/ml were observed after oral doses of 15, 30, 60 and 120 mg, respectively. These peak concentrations were achieved within 30–60 min from oral administration.

In acute therapy peak levels of 6.7, 11.1, 16.5 and 23.5 ng/ml were achieved after the patients received doses of 15, 30, 60 and 120 mg, respectively. The area under the concentration time curve was greater during chronic than during acute treatment.

A marked interpatient variability in plasma concentrations of ISDN after any given oral dose is found. This may be attributed in part to the low systemic bioavailability of the drug.

25.2.2. Distribution

Large distribution volumes of 80–470 L were calculated in human subjects receiving ISDN intravenously. The distribution of intravenously administered ISDN is rapid, with a half-life of 5 min [384]. Following oral administration a volume of distribution of 600 L has been determined [385]. This large volume, exceeding the body compartment volumes, indicates that the drug is extensively distributed in the tissues. The drug accumulates in the heart, vascular walls, lungs, kidney and liver, and particularly in fatty tissues.

About 30% of circulating ISDN is protein-bound [386].

25.2.3. Metabolism, Half-Life and Elimination

ISDN undergoes extensive first-pass hepatic metabolism. It is rapidly denitrated by the enzyme glutathione *S*-transferase during the first-pass in the liver and only a small amount reaches the systemic circulation. The systemic bioavailability shows marked interpatient variability: after oral administration it was found to range between 1% and 75%, after sublingual administration from 19% to 93% and after topical administration from 10% to 60% [252, 253]. The systemic bioavailability of slow-release ISDN was found to be about 90% of that of the conventional formulation [387].

The elimination half-life of ISDN was found to range between 0.5 and 10.0 h. In most studies half-life was 30–60 min. It was similar for oral and sublingual ISDN. It has also been described as biexponential, with an elimination half-life of about 1.5 h in the alpha phase and about 4 h in the beta phase [386]. The elimination half-life during chronic treatment may be longer than that observed after acute administration, probably due to saturation of the metabolic degradation system.

Taylor et al [388] observed a biexponential decline of ISDN plasma concentration after discontinuation of a 2-h intravenous infusion of ISDN. The half-lives of the alpha and beta phases were 10 min and 80 min, respectively. Straehl and Galeazzi [384] found a biexponential decline of plasma concentration of ISDN following intravenous administration of 5 mg over 5 min. The half-life alpha phase was 4.7 min and beta phase 54.7 min. Half-lives of the alpha and beta phases following sublingual administration of 10 mg of ISDN were 8.7 and 48.8 min, respectively. These phases represent rapid distribution and early elimination.

The systemic clearance of ISDN after intravenous administration varies from 1.5 to 4.0 L/min. These values are higher than the combined renal and hepatic blood flows. This indicates metabolism of the drug in other organs. Some of the differences may be attributed to metabolism in vascular walls.

The duration of action of ISDN given orally in the form of conventional tablets is about 8 h in angina pectoris and 3.5–5 h in congestive heart failure during acute treatment. During long-term treatment the duration of action in both conditions may be shortened to 2 h. This depends not only on the plasma half-life but also on the interaction of the drug with vascular receptors and the possible development of vascular tolerance and/or other forms of tolerance.

ISDN is eliminated by hepatic denitration. The isomeric mononitrate metabolites, IS-5-MN and IS-2-MN, persist in the plasma longer than the parent drug. These mononitrates are denitrated slowly and possess vasodilator properties which are qualitatively similar to those of ISDN. They are discussed in a separate chapter, and their clinical pharmacology is discussed in the next section of this chapter.

ISDN or its metabolites do not inhibit or induce metabolism during chronic administration [389]. Approximately 10% of on oral dose of ISDN is excreted in the urine as conjugated ISDN and 1% as IS-2-MN. Most of the amount is excreted as isosorbide.

The fact that orally administered ISDN is effective despite the extensive first-pass hepatic metabolism may be explained, at least in part, by the activity of mononitrates. Saturation of the metabolic pathway, and predominantly the nitrate reductase hepatic system, may also account for the activity. The high doses of oral ISDN which are required to achieve an effect (compared with the low doses of sublingual nitroglycerin and ISDN) may be required to saturate this pathway.

Chewable ISDN has a more rapid onset of action than the oral formulation. The effect is evident within 10 min from administration and peak activity within 25–30 min. The mean duration of action is about 2 to 2.5 h, a period shorter than that of orally administered ISDN. Therefore the chewable drug is more useful for relief of pain than for prophylaxis in patients with angina pectoris.

ISDN can also be given intravenously. By this route the onset of action is very rapid, within a few minutes, and steady state is reached within about 30 min. Another important advantage of this method is the relatively constant plasma concentration, without marked fluctuations, which is achieved by it. Dose titration is also most easily achieved by this method. The possibility of continuing oral treatment with the same drug administered intravenously is an important advantage.

Recently an interesting view of redistribution of ISDN has been suggested. The in vitro half-life of ISDN in human blood was found to be about 1.5 h – almost five times shorter than the mean in vivo half-life during chronic treatment. This finding indicates that the drug may accumalate in tissues and then redistribute, returning to the plasma and being eliminated.

The pharmacokinetic parameters of ISDN are not considerably different between patients with congestive heart failure and those with angina pectoris, between smokers and nonsmokers and between fasting and postprandial conditions. Two minor differences, however, were observed by Fung et al [390]. There was a trend for higher plasma levels of ISDN in smokers and for food to reduce the apparent absorption rate of the drug.

As ISDN undergoes first-pass hepatic metabolism, patients with impaired hepatic metabolism may have higher plasma concentrations of this drug [391]. Bogaert et al [392] reported that in five out of seven patients with cirrhosis of the liver and extensive shunting, plasma concentrations of ISDN after oral administration (10 mg) were markedly higher than those found in subjects without hepatic disease. This was attributed to a higher systemic bioavailability and/or a lower systemic clearance of the drug due to shunting and hepatic cell dysfunction. The difference was confirmed by the calculation of the areas under the concentration time curve. These findings indicate that the dosage of ISDN should be reduced in patients with severe hepatic dysfunction.

25.3. IS-5-MN

IS-5-MN is given intravenously or orally. Other routes of administration, as used for other nitrates, may theoretically be used also for IS-5-MN. However, there is no need to give such a long-acting drug by the sublingual route or, on the other hand, by the transdermal route. IS-5-MN is rapidly and completely absorbed from the gastrointestinal tract after oral administration.

IS-5-MN has complete absolute bioavailability due to the absence of any first-pass hepatic or intestinal effect [286].

Biamino et al [305] reported that the hemodynamic effects of 40 mg IS-5-MN were evident at 15–20 min after oral administration and reached their peak after 45 min.

IS-5-MN follows dose-linear kinetics after single as well as multiple doses. Kenedi [301] reported a linear increase of plasma concentrations with the increase in dose of IS-5-MN (slow-release capsules): 235 ± 57 ng/ml with 25 mg, 492 ± 103 ng/ml with 50 mg and 993 ± 197 ng/ml with 100 mg. Measurements were made 8 h after dosing. Abshagen [393] reported an excellent linear relationship between the doses of IS-5-MN and the respective area under the concentration curve after single doses.

These investigators also evaluated the kinetics of IS-5-MN, 20 mg orally, three times daily for 1 week, in 18 patients with coronary artery disease. The resulting area under the concentration time curve within one dosage interval during steady state was not different from the total area under the concentration time curve after a single oral dose of 20 mg in healthy volunteers. Thus IS-5-MN follows dose-linear kinetics also after multiple dosing.

IS-5-MN does not undergo any first-pass hepatic degradation. This was confirmed by the finding of similar plasma concentrations after oral and intravenous administration of 20 mg IS-5-MN to healthy volunteers [393].

After intravenous administration plasma levels of IS-5-MN and IS-2-MN decline biexponentially and may be described by an open, two-compartment body model. Distribution half-life of IS-5-MN was found to be 8.6 min and that of IS-2-MN 12.5 min [394].

The volume of distribution measured after intravenous administration of IS-5-MN was 0.6 L/kg [393]. In another study the volume of distribution of intravenously administered IS-5-MN in healthy volunteers was 48 L and that of IS-2-MN 55 L [394]. In another study the volume of distribution was 48.5 L [387].

Plasma concentrations of IS-5-MN show little variation, of no more than 25% [395], as compared with variations of up to 75% with ISDN. Strong correlation was found between the plasma concentration of IS-5-MN and the effect. For example, Reifart et al [396] observed excellent correlation of the time curve of IS-5-MN concentrations and the decrease in pulmonary capillary wedge pressure in patients with acute myocardial infarction.

IS-5-MN is eliminated by denitration and glucuronidation.

In a study in healthy volunteers only about 21% of the orally administered dose was excreted unchanged in the urine, 30% was excreted as isosorbide and 17% as the 2-glucuronide of IS-5-MN. Neither of these metabolites is pharmacologically active.

A high renal clearance of the glucuronide (about 1.8 L/min), exceeding renal plasma flow, has been found in healthy volunteers, suggesting the occurrence of substantial intrarenal glucuronidation of IS-5-MN. Thus the kidneys are a major site of formation of this metabolite.

The metabolic rate of break-down declines from ISDN to IS-2-MN to IS-5-MN due to steric protection of the nitro group in the 5-endo position. This reduced elimination rate compensates for the respective decrease in vasodilating activity. The elimination half-life of IS-5-MN is longer than that of all other organic nitrates. After intravenous infusion of 20 mg (to healthy volunteers) the elimination half-life was 4.6 h. In another group of healthy volunteers the elimination half-life of IS-5-MN was 4.15 h and the half-life of IS-2-MN was 1.9 h [394]. In another study in healthy volunteers the elimination half-life of IS-5-MN was 4.4 h [387].

The duration of action of a conventional tablet of IS-5-MN is about 6 h. A slow-release formulation can prolong it to about 8.6 h. In an investigational formulation there is a conventional portion which provides rapidly about 40% of the dose and at a slower rate the remaining portion.

The elimination half-lives of the metabolites isosorbide and the 2-glucuronide of IS-5-MN are 8 and 6 h, respectively.

The clearance of IS-5-MN, measured after intravenous administration of 20 mg, was 115 ml/min. This was made up almost exclusively by the extrarenal route [397]. In another study in healthy volunteers total plasma clearance of IS-5-MN was 8.5 L/h after intravenous administration of 5 mg. Total plasma clearance of IS-2-MN after a similar dose was 23.2 L/h [394].

IS-5-MN is not eliminated by peritoneal dialysis to any significant extent. ISDN undergoes first-pass hepatic metabolism. Therefore it may accumulate in patients with hepatic dysfunction. As IS-5-MN does not undergo a first-pass hepatic metabolism, such accumulation may not take place. Akpan et al [398] studied the metabolism of IS-5-MN in six patients with liver cirrhosis and decreased antipyrine clearance. They found that kinetic parameters, including area under the concentration curve, half-life, clearance and volume of distribution were not different from those measured in healthy human subjects. This indicates the absence of any significant presystemic metabolism of the drug.

Renal clearance of the metabolite IS-5-MN 2-glucuronide was markedly reduced in the patients with hepatic cirrhosis, despite normal renal function. There is no explanation for this finding, but it has no clinical relevance because the glucuronide is pharmacologically inactive.

The dosage regimen of IS-5-MN should not be changed in patients with impairment of liver function.

The study of Akpan et al [398] included also a patient in whom a portocaval shunt had been performed. Even in this patient no pharmacokinetic differences were observed. Mannebach et al [399] studied the pharmacokinetics of IS-5-MN in elderly patients and found no differences as compared with young healthy volunteers. As IS-5-MN has a low renal clearance and insignificant urinary excretion, it is not accumulated even in patients with severe renal failure. Raue et al [400] studied the pharmacokinetics of IS-5-MN, 20 mg twice daily for 7 days, in five patients with severe renal failure (creatinine clearance 10–18 ml/min). The elimination half-life did not change after multiple doses, indicating that the metabolism of the drug was not impaired.

The two primary metabolites of IS-5-MN in this group of patients were isosorbide and IS-5-MN 2-glucuronide. Renal clearance of the glucuronide was found in these patients to be only about 5 times higher than the glomerular filtration rate, which is considerably lower than in human subjects with normal renal function. These findings indicate that capacity for tubular secretion of the glucuronide is impaired but not abolished in patients with renal dysfunction. The glucuronide accumulated in the plasma. As it does not have significant vascular activity, no dose adjustment of IS-5-MN is required in patients with renal failure [400]. Excretion of isosorbide is also impaired in renal failure, but its accumulation has no pharmacodynamic significance.

In summary, IS-5-MN shows a more consistent and more predictable pharmacokinetic profile after single and multiple dosing in patients of all ages, including those with hepatic or renal dysfunction.

Tanaka et al [324] compared the pharmacokinetics of IS-5-MN after oral administration of IS-5-MN and ISDN in patients with chronic congestive heart failure. The maximal plasma concentration of IS-5-MN was 779.68 ± 228.74 ng/ml after administration of this drug and only 97.76 ± 27.89 ng/ml after administration of ISDN. The elimination half-life was 5.60 ± 2.58 h and 2.3 ± 1.3 h, respectively.

Davies and Williams [401] evaluated the pharmacokinetics of IS-5-MN after chronic oral doses of two formulations: IS-5-MN tablets versus IS-5-MN slow-release capsules in young healthy volunteers. The slow-release properties of IS-5-MN capsules were shown by a median time to maximal plasma levels of 5 h, and a median time above a plasma level of 100 ng/ml of 17 h. Despite a higher dose of the tablets the mean maximal plasma level was not significantly different between the formulations. As might have been expected, the bioavailability of IS-5-MN given as slow-release capsules was lower than that of the tablets (84% versus 100%, $p < 0.005$). Elimination kinetics and tolerance were similar for both formulations.

No drug accumulation was observed after 14 days of continuous administration of IS-5-MN in patients with renal insufficiency [401a].

25.4. IS-2-MN

IS-2-MN is not used clinically and its pharmacokinetic profile will therefore not be discussed in detail. It is eliminated by denitration or glucuronidation.

25.5. Metabolism of IS-5-MN Compared with ISDN – Toxicologic View

ISDN is rapidly and almost completely denitrated in the liver, forming IS-5-MN and IS-2-MN. IS-5-MN is also denitrated but at a much slower rate than ISDN. Therefore administration of ISDN results in greater and more rapid formation of nitrite than IS-5-MN. The nitrite ion is responsible for an important toxicologic problem associated with nitrate therapy, namely methemoglobinemia.

Strein et al [402] evaluated the metabolism of both these compounds in dogs. They found that at equimolar doses IS-5-MN caused either no significant rise or one-tenth the rise of the nitrite ion produced by ISDN. ISDN produced an increase in serum alkaline phosphatase which was evident 15 days after initiation of treatment. No such increase was observed with IS-5-MN.

25.6. *Metabolism of Nitrates in Vascular Walls*

Unlike most other drugs, nitrates, and particularly nitroglycerin, are metabolized not only in the liver but also in the walls of blood vessels [403]. This metabolic process is probably related to the mechanism of action of nitrates. It results in considerable arteriovenous differences in plasma concentration of nitroglycerin. Imhof et al [404] have recently reported that the concentration of nitroglycerin in arterial blood is about 3 times higher than in venous blood, although this arteriovenous difference is subject to marked interpatient variability.

A first-pass metabolic process in vascular walls has been directly demonstrated after intravenous injection of nitroglycerin [403]. The concentration ratio between the vascular wall and plasma decreased with increasing distance of the sampling site from the point of injection. This ratio increases in direct relation with time from administration.

After nitrates are discontinued the rate of disappearance of the pharmacodynamic effects is slower than the rate of decline of the plasma concentrations of the drug [405]. This may be attributed to slower clearance of nitrates from the vascular walls than from the plasma.

25.7. *Relationship Between the Pharmacokinetic Effect of Nitrates and Their Therapeutic Effect*

Evaluation of the relationship between pharmacokinetics and pharmacodynamics of nitrates reveals that the plasma concentration of nitrates is not directly related to the therapeutic effect. It is difficult to describe the pharmacological effects of nitrates as a dose-response or concentration-response relationship.

This results from several factors:

1. The end-point therapeutic effect of nitrates is a secondary effect, resulting from different effects in various segments of the circulation (described in detail in the chapter on the mechanism of action of nitrates). In each segment there is a different concentration-response curve. Moreover, the relative contribution of each effect (coronary dilation, arterial dilation and venodilation) is variable and differs in various patients and conditions.
2. Nitroglycerin is metabolized also in vascular walls and this process is related to the mechanism of action of the drug. It has been suggested that the pharmacodynamic effects of nitroglycerin may correlate better with vascular wall concentration than with plasma concentration [403].
3. The sensitivity of vascular walls to nitrates show marked intersubject variability.
4. Much of the pharmacodynamic effect of ISDN is exerted by its active metabolites, IS-5-MN and IS-2-MN. These metabolites are cleared more slowly than the parent drug and also complicate the concentration-response relationship. The situation is further complicated by interference of the mononitrates with the metabolism of the parent drug.
5. The concentration-response relationship is further complicated by the development of tolerance, even if basically the pharmacodynamic effect is directly related to concentration, as is tolerance. Therefore the concentration-re-

sponse curve might differ between acute and chronic administration of nitrates. As the various nitrates have different degrees of tolerance, the influence of tolerance on the concentration-response curve should be individually determined for each agent.
6. Tolerance to some of the vascular effects of nitrates may develop in the presence of preserved antianginal activity.
7. Other pharmacokinetic parameters, such as the rate and direction of changes in plasma concentration of nitrates may also affect the pharmacodynamics of these drugs.
8. Counterregulatory mechanisms, including activation of adrenergic activity, the renin-angiotensin system and other mechanisms attenuate the initial effect of nitrates.

All these factors complicate the dose-response and concentration-response relationship of nitrates.

This is only partially valid for IS-5-MN, which has a much simpler pharmacokinetic profile than ISDN. IS-5-MN shows a close correlation between plasma concentration and pharmacodynamic characteristics in patients with angina pectoris [406] and in patients with recent myocardial infarction [396]. Nevertheless, certain correlations between concentration and response can also be made for ISDN. Peak hemodynamic effect and maximal plasma concentration usually occur at approximately the same time [388]. The pharmacodynamic effect is then sustained while the plasma-concentration declines. This may be attributed to the effect of mononitrates. Dose-linear kinetics have been reported after single oral administration of ISDN. At least one study revealed a dose-linear relationship also during chronic therapy [407].

25.8. Discontinuation of Nitrates

Abrupt discontinuation of nitrates may be associated with withdrawal phenomena, including myocardial infarction, angina pectoris, silent ischemia, hypertension and aggravation of congestive heart failure. However, withdrawal phenomena, particularly those associated with coronary artery disease, are very uncommon with nitrates. When antianginal drugs are discontinued (prior to exercise test, etc.) it is common practice to discontinue calcium antagonists and beta-blockers over several days and to continue nitrates until the last day, when they are withdrawn or tapered-off over 1–2 days.

Significant signs and symptoms of nitroglycerin withdrawal have been reported only after cessation of massive industrial exposure and not after therapeutic use in patients with coronary artery disease. However, in some patients with congestive heart failure due to coronary artery disease deterioration of the clinical condition has been observed after abrupt discontinuation of nitrates.

Abrupt discontinuation of intravenously administered nitrates in patients with acute myocardial infarction or unstable angina is absolutely contra-indicated except when shock develops.

Rudolph et al, who studied extensively the question of nitrate tolerance and intermittent therapy have studied also the related question of rebound phenomena after withdrawal of nitrates. In a recent study [407a] they evaluated exercise-induced electrocardiographic changes before and after termination of a nitroglycerin infusion in 20 patients with coronary artery disease. No clinically-relevant rebound phenomena have been observed after termination of the nitroglycerin infusion.

26. Side Effects and Drug Interactions

26.1. Side Effects

The side effects of nitrates include general effects common to most vasodilators and specific effects unique to nitrates. The general effects derive from the same properties which are responsible for the therapeutic effect of nitrates, mainly peripheral vasodilation. In addition, the alcoholic diluents of some intravenous nitrates may also cause intoxication.

Adverse effects attributed to vasodilation include headache, hypotension, postural hypotension, impaired coronary and cerebral perfusion, dizziness, syncope, weakness, flushing and palpitations.

Headache is the most common adverse effect of nitrates. It appears in most patients in the first days after initiation of long-term nitrate therapy. Its incidence with sublingual nitroglycerin is difficult to estimate, but it has been suggested to range between 10% and 90%. Headache is the major cause of discontinuation of treatment. It usually disappears with continued treatment but may require reduction in dose or use of an analgetic agent.

Postural symptoms may occur in patients both with and without congestive heart failure but they are much more common in those with pretreatment and normal left ventricular filling pressure. They result from the combined effect of nitrates on left ventricular filling pressure and on systemic vascular resistance. The so-called "nitrate syncope" is often observed in emergency rooms. It usually develops in patients with angina pectoris who take more than one sublingual tablet at very short intervals without waiting for the drug to exert its effect. Nitrate-induced orthostatic hypotension is uncommonly observed during chronic treatment. Even when it develops at initiation of therapy, it usually persists for no longer than 10 h. There is a rapid adaptation of blood pressure regulation to chronic nitrate therapy (which is not related to potential tolerance in the venous system).

The most important adverse effect resulting from the hemodynamic effects of nitrates is decrease in coronary perfusion pressure due to decrease in systemic arterial pressure. The decrease in coronary perfusion pressure may aggravate myocardial ischemia. This potential risk is seldom expressed in patients with coronary artery diseases who are on long-term oral nitrate therapy. It is more significant in patients treated with intravenous nitrates in coronary care units. However, appropriate monitoring minimizes this risk. There is no study demonstrating differences among the various oral formulations in the frequency of adverse effects resulting from vasodilation.

Increased sensitivity to the hypotensive effect of nitrates was observes in 31% of the patients with acute myocardial infarction in one recent study [147a]. Seventy-two percent of them had an inferior infarction. Moreover, in the 9% of the patients whose mean blood pressure decreased to less than 80 mmHg between 3 and 12 h of the nitroglycerin therapy, it results in larger infarcts compared with placebo and a smaller beneficial effect on left ventricular function and topography compared with other patients receiving nitroglycerin in whom excessive lowering of blood pressure was ot observed.

Up to 60% of patients with acute inferior wall myocardial infarction develop hypotension. A marked hypotensive response to nitrates may suggest the presence of right ventricular infarction. As hypotension could be anticipated after administration of nitrates to patients with inferior wall infarction and right ventricular involvement, nitrates should be used carefully in such patients [408a]. The mechanism of hypotension in these cases is, mainly, direct venodilation resulting in a decrease in ventricular filling and cardiac output. This is of particular significance in patients with right ventricular infarction who require high filling pressures to maintain their cardiac output.

Another study revealed, however, that severe ISDN-induced hypotension in patients with acute myocardial infarction is more prevalent in anterior infarction [408b]. It could not be clearly predicted from resting hemodynamic or blood volume measurements, at least in non-hypotensive patients. The suggested mechanism was marked venous and arterial vasodilation, probably during a stage of a severe impairment in ventricular compliance.

Several uncommon specific side effects of nitrates have been observed; these are discussed below.

26.2. Hypersensitivity

Hypersensitivity to nitroglycerin is a very rare, poorly defined and poorly understood condition. It has been observed with small doses of the drug. The reaction consists of bradycardia and circulatory collapse. Intravenous atropine is effective in management of this condition [409, 410].

The potential role of this adverse reaction has been emphasized by a recent report by von Armin et al [411] of a 62-year-old man who developed an inferior wall myocardial infarction during ambulatory ST-segment monitoring. Before the onset of persistent ST-segment elevation the patient had taken several tablets of nitroglycerin sublingually. This was followed by bradycardia and symptoms of hypotension. This could have resulted from a paradoxical reaction to nitroglycerin. The bradycardia-hypotensive response may also be found in patients treated with high doses of nitrates intravenously. Page et al [72] reported two cases of this syndrome among 67 patients with unstable angina treated with high doses of nitroglycerin intravenously. It disappeared after diminution of the dose. Come and Pitt [410] reported a higher incidence of severe hypotension and bradycardia in patients with acute myocardial infarction treated with nitroglycerin. They found seven episodes of hypotension and absolute or relative bradycardia in five patients receiving either sublingual nitroglycerin (two patients) or intravenous nitroglycerin (three patients) within 24 h of the onset of chest pain. Four of these were hemodynamically monitored, and showed reductions in left ventricular filling pressure simultaneously with the hypotension-bradycardia event. No initial increase in heart rate was observed in any of the patients before the appearance of bradycardia.

The lower incidence of the bradycardia-hypotension syndrome in the patients reported by Page et al [72] may result from the slow titration of nitroglycerin infusion rate, which allowed (presumably) physiologic adaptation.

As early as 1932 Prodger and Ayman [412] observed four cases of bradycardia and hypotension in a group of 110 patients with hypertension who received sublingual nitroglycerin. Other investigators reported that tilting after sublingual administration of nitroglycerin resulted, in some patients, in hypotension, bradycardia and syncope [413, 414]. An initial increase in heart rate has been frequently observed in these cases. The mechanism of bradycardia associated with nitrate-induced hypotension is not known. Careful observation is required to detect and treat it early, particularly in patients with acute myocardial infarction and in those receiving high doses of nitrates intravenously.

26.3. Increased Methemoglobin Levels

Nitrates can oxidize the iron in the hemoglobin molecule, from the ferrous to the ferric state, and form methemoglobin, which can not transport oxygen. Elevated concentrations of methemoglobin were found after administration of therapeutic doses of nitroglycerin [415].

This adverse effect may be potentially deleterious in two groups of patients:

1. Infants, in whom nitrates are rapidly transformed to nitrites in the gastrointestinal tract. The nitrites, which are very toxic for hemoglobin, are rapidly absorbed from the gastrointestinal tract. As nitroglycerin is now used in infants and children with hemodynamic impairment after cardiac surgery this potential risk should be remembered.
2. In patients with coronary artery disease methemoglobinemia might have deleterious effects on the myocardial oxygen supply/demand ratio. This potential risk, however, is of little relevance in clinical practice. It is more important in patients who receive high doses of nitroglycerin intravenously. Gibson et al [416] reported that infusion rates higher than 7 μg/kg per minute were associated with methemoglobinemia. This complication should be considered in any patient treated with this drug who appears cyanotic in the presence of normal arterial oxygen saturation. Methemoglobin levels higher than 3% indicate toxicity. Patients with hepatic disfunction are more susceptible to methemoglobinemia.

26.4. Ventricular Fibrillation

There is only one report, by Daskoff et al [417], stating that intravenous administration of nitroglycerin resulted in a significant increase in the ventricular fibrillation threshold. Other investigators, using a similar protocol, were unable to demonstrate any effects of nitroglycerin on ventricular fibrillation threshold in normal hearts.

26.5. Prolongation of Neuromuscular Blockade

Glisson et al [418] reported prolongation of pancuronium-induced neuromuscular blockade by intravenously administered nitroglycerin. This effect may be attributed to a specific interaction between nitroglycerin and pancuronium, as it has not been observed with other neuromuscular blockers.

26.6. Elevation of Intracranial Pressure

Rogers et al [419] reported that nitroglycerin may further increase elevated intracranial pressure. Although no other study confirming this finding has been reported, it is recommended not to give nitroglycerin intravenously to patients with elevated intracranial pressure.

26.7. Ethanol and Propylene Glycol Intoxication

Although the hemodynamic effects of intravenous nitroglycerin are carefully monitored during administration, no attention is usually given to the effects of the drug's diluent. Shook et al [420] have recently reported two cases of alcohol intoxication in patients receiving high doses of intravenous nitroglycerin. These cases emphasize the advantage of using nitroglycerin in nonalcoholic solution. Such formulation is now available.

Thiamine serves as a cofactor for glucose metabolism in the glycolytic and pentose phosphate pathways. Its deficiency produces a nutritional neurologic disorder, the Wernicke-Korsakoff syndrome. Use of ethyl alcohol is commonly associated with thiamine difficiency, both because ethanol metabolism inhibits thiamine pyrophosphorylation, and because the use of ethanol is associated with nutritional deficiency. Recently a case of sudden onset of Wernicke's encephalopathy was described in a nonalcoholic patient with angina pectoris. The encephalopathy was produced by the ethanol and propylene glycol diluents used to dissolve intravenous nitroglycerin with which the patient was treated [421].

Adverse effects of nitrates in the elderly are similar, qualitatively and quantitatively, to those observed in younger patients.

26.8. Drug Interactions

Nitrates are involved in fewer true drug interactions than any other antianginal agent. Even the interactions that do occur are minimal and of little clinical relevance.

26.8.1. Aspirin

Weber et al [422] reported that pretreatment with 1 g aspirin increased plasma levels of nitroglycerin by about 50%. The pharmacodynamic effect of nitroglycerin had concomitantly increased. This interaction has been attributed to a reduction in hepatic blood flow by aspirin, resulting in a decrease in hepatic clearance of nitroglycerin.

Aspirin may theoretically (similarly to other inhibitors of prostaglandin synthesis) attenuate the effect of nitroglycerin, if it is accepted that nitrates exert their effect via the release of prostaglandins. No report of such interaction has been published.

Simonetti et al [423] performed a study aiming to determine whether coronary vasodilation by nitrates in man is mediated by prostaglandins. They studied the effect of ISDN on coronary artery diameter and coronary blood flow, before and after acute administration of aspirin, in 13 patients, seven of them with normal coronary arteries and four with coronary artery disease. The effects of intracoronary ISDN on coronary artery diameter and coronary blood flow were studied by cardiac catheterization and angiography, before and after intracoronary and intravenous administration of aspirin, 100 mg and 1 g, respectively. Aspirin did not produce any significant changes in coronary artery diameter and did not alter the response to ISDN. It did not prevent the decrease in coronary resistance produced by ISDN. It may be concluded that coronary vasodilation by nitrates is not mediated by prostaglandins and can not be prevented by aspirin.

26.8.2. Indomethacin

Indomethacin is an inhibitor of prostaglandin synthesis. By this mechanism it may impair the vasodilatory effect of nitrates. This was shown by Van Dusen and Fischi [424] with nitroglycerin. Other studies, however, did not confirm such an interaction. To clarify this point Thadani and Kellerman [425] studied the influence of indomethacin on the circulatory and antianginal effects of nitroglycerin in six patients with chronic stable angina pectoris. The patients were treated with indomethacin, 50 mg three times daily, or placebo, for 1 week each in a double-blind, randomized protocol. Heart rate, systemic arterial pressure and ST-segment depression were measured, at rest and during exercise, before and after the administration of 0.6 mg sublingual nitroglycerin during indomethacin and placebo therapy.

Nitroglycerin decreased standing systolic blood pressure by 38 mmHg during placebo treatment and by 36 mmHg during indomethacin treatment. Heart rate increased to the same extent during placebo and indomethacin therapy. During placebo nitroglycerin increased the duration of exercise to onset of angina by 128 sec and during indomethacin by 84 sec (difference nonsignificant). The increased total duration of exercise was similar during placebo and indomethacin. No significant difference was observed between the reduction in ST-segment depression by nitroglycerin during placebo and indomethacin.

These findings show that indomethacin did not attenuate the vascular and antianginal effects of nitroglycerin. It is most probable, therefore, that these effects are not mediated by prostaglandins.

26.8.3. Barbiturates

Barbiturates are hepatic enzyme inducers. Concomitant administration with nitrates may result in accelerated hepatic metabolism of the nitrates, with lower plasma concentrations.

26.8.4. Vasodilators and Other Antianginal Drugs

There is no true interaction between nitrates and other vasodilators or other antianginal drugs such as beta-adrenoreceptor blockers. However, nitrates potentiate the effects of all these agents. The problem of nitroglycerin-induced orthostatic hypotension may be more pronounced in patients concomitantly treated with other drugs that tend to cause hypotension. This is often the cause of the "nitroglycerin syncope" – fainting of patients with angina pectoris who are treated with calcium antagonists, beta-blockers or long-acting nitrates after they have taken nitroglycerin sublingually. In patients treated with beta-blockers the increase in heart rate in response to nitroglycerin-induced vasodilation is inhibited and orthostatic hypotension may develop [426].

26.8.5. Heparin

It has been reported that intravenously administered nitroglycerin may interfere with the anticoagulant effect of heparin. This effect was attributed to the diluent propylene glycol, which is used in some of the intravenous formulations of nitroglycerin. To evaluate this interaction, Habbab et al [427] studied eight infusions of nitroglycerin and heparin in seven patients, by monitoring of the partial thromboplastin time (PTT). In six of the infusions nitroglycerin with propylene glycol was used, and in the remaining two infusions nitroglycerin without propylene glycol.

In all eight infusions, regardless of whether the formulation contained propylene glycol, the response was similar. The dose of heparin required to prolong the PTT to twice the control level in patients on nitroglycerin was higher than the usual dose, and additional intravenous bolus injections of heparin were required to maintain therapeutic PTT values. The PTT was shortened, despite a constant heparin infusion, whenever the nitroglycerin infusion rate was increased. It was prolonged when the nitroglycerin infusion rate was decreased. On discontinuation of nitroglycerin there was a rebound overshoot of the PTT, in the presence of a constant infusion rate of heparin. In six of the eight patients, including the two patients who received nitroglycerin without propylene glycol, the PTT increased to more than 3.5 times the control level.

In a larger study Pizzulli et al [427a] evaluated the interaction between heparin and nitroglycerin in 27 patients [427a]. They found that intravenous administration of nitroglycerin induces heparin resistance, without altering the plasma levels of heparin.

These findings confirm that intravenous administration of nitroglycerin inhibits the anticoagulant effect of heparin. This does not result from the diluent propylene glycol but from the nitroglycerin itself. Hence, in patients who receive both intravenous nitroglycerin and heparin, resistance to the anticoagulant effect of heparin might be expected. In these patients the PTT should be often monitored and high doses of heparin should be used. On discontinuation of nitroglycerin the infusion rate of heparin should be lowered to avoid rebound effects of heparin [428].

26.8.6. Other Drugs

No interaction has been observed between nitrates and digitalis, glycosides, diuretics and other drugs commonly used in cardiovascular diseases.

27. Dosage and Administration

27.1. Nitroglycerin

27.1.1. Sublingual

Doses of sublingual tablets of nitroglycerin usually range between 0.3 and 0.8 mg. This dose may be repeated two to three times, at intervals of 3–5 min, but only if pain is not adequately relieved. Patients should be carefully instructed on this point, as very often patients, particularly those for whom the disease and drug are new, tend to take several subsequent tablets without waiting for the therapeutic effect. This may lead to adverse effects. On the other hand, patients with chronic stable angina pectoris who become familiar with the drug find quite rapidly their optimal dose and do not exceed it.

Sublingual nitroglycerin tablets may be used by patients on long-term nitrate therapy who develop acute episodes of angina pectoris.

27.1.2. Nitroglycerin Oral Spray

Several manufacturers produce nitroglycerin oral spray. Individual description of these is beyond the scope of this book. The usual dose is 1–3 squirts, at intervals of 2–5 min.

27.1.3. Intravenous

The infusion rates of nitroglycerin should be individually adjusted to each condition and each patient. In congestive heart failure it should be preferably given under hemodynamic monitoring and titrated according to left ventricular filling pressure and systemic arterial pressure and resistance.

The initial infusion rate should not exceed 1 mg/h. This dose is gradually increased, every 1–3 h. Until a few years ago infusion rates of not more than 6 mg/h were used. Recently in many centers infusion rates of up to 15 mg/h have been effectively and safely used. Although some investigators have experience with higher doses, it is not recommended at present to exceed 15 mg/h. A practical approach in patients with severe angina pectoris is to initiate treatment with a dose of 10 μg/min, increase it by 10 μg/min every 5 min up to 50 μg/min, and then to further increase it by 50 μg/min after each anginal episode.

Recently Zimrin et al [382] reported a dose-finding study to evaluate the maximal tolerated dose of intravenous nitroglycerin. Titration began at 10 μg/min with six increments of 10 μg/min followed by six increments of 20 μg/min. Maximal tolerated infusion rate (which decreased blood pressure to 100 mmHg) ranged from 10–120 μg/min (mean 59 μg/min), or 0.70 ± 0.45 μg/kg per minute.

Discontinuation of infusion should be performed gradually. If intravenous nitroglycerin should be replaced by oral nitrates, oral therapy should start a few hours before discontinuation of the infusion.

27.1.4. Intracoronary

Intracoronary nitroglycerin is used for investigational purposes, for relief of coronary spasm and as part of interventional procedures such as thrombolysis and percutaneous transluminal coronary angioplasty. Most investigators have used a bolus injection of 25–1150 μg, injected over 15–45 sec. Others used a 50 μg/min dose at a constant infusion over 3 min. A dose of about 200 μg is probably optimal for near-maximal coronary dilation without clinically important systemic hemodynamic effects. With doses higher than 300 μg systemic effects may be observed.

27.1.5. Transdermal

Although nitroglycerin ointment is available, the only application widely used clinically is transdermal nitroglycerin patches. The commercially available patches deliver 5–10 mg over 24 h. Doses of 5–60 mg/24 h have been used. It is recommended to use doses in the lower range to avoid development of tolerance. The 5 mg/24 h dose is probably too low. It is recommended to use the 10 mg/24 h dose for most cases in clinical practice, particularly for patients with angina pectoris. Higher doses may be used in resistant cases. Most patients respond to doses of up to 20 mg/24 h. Black patients may require higher doses because the rate of absorption and total quantity absorbed are less in black patients than in Caucasian.

27.1.6. Oral Slow Release

Use of orally administered nitroglycerin is controversial. The dose should be high enough to achieve therapeutic plasma levels despite the extensive first-pass hepatic metabolism. Doses of 2.5–6.5 mg are used.

27.1.7. Buccal Nitroglycerin

Buccal nitroglycerin tablets are placed in the buccal cavity between the teeth and the upper lip and adhere to the mucosa. Doses of 1–6.5 mg were used to treat angina pectoris and doses of 10 mg three times daily for congestive heart failure. The initial dose should be 1 mg only, to avoid hypotension. A twice-daily administration may be sufficient for patients who maintain the tablet for at least 5 h. Patients who maintain the tablet for less than 3 h may require four daily doses.

27.1.8. Pediatric Dose

In infants and children nitroglycerin is given intravenously at infusion rates of about 5 μg/kg per minute.

27.2. ISDN

27.2.1. Sublingual

The conventional medication of ISDN for relief of pain is a sublingual tablet. Doses of 2.5–5 mg are generally used. The dose may be repeated once or twice within a few minutes. It is not recommended to take more than 10 mg over 30 min. In patients with systolic arterial pressure of 95–105 mmHg the initial dose should be 1.25 mg.

27.2.2. Oral

Oral administration of ISDN is the usual route of administration for prophylaxis of angina. Most patients receive a dose of 10–40 mg, three to four times daily. In severe cases doses of 20 mg are given every 4 h or doses of 10–20 mg are given every 2 h. Some investigators recommend twice-daily administration to prevent tolerance.

Slow-release oral formulations are available for once-daily administration. Sublingual nitrates may be used in patients on long-term oral nitrate therapy who develop acute episodes of angina pectoris.

27.2.3. ISDN Oral Spray

One squirt of ISDN oral spray containing 1.25 mg, it is recommended to start with one squirt and repeat it once or twice after 3–5 min if the response to the first dose is inadequate. It is not recommended to give more than three squirts in 10 min. ISDN oral spray may be used in patients on long-term nitrate therapy who develop acute anginal episodes.

27.2.4. Intravenous

Infusion rates of 1–15 mg/h are used in most centers. It is recommended to start with 1 mg/h and increase the rate, if required, every 30–60 min. More rapid dose adjustment may be required in severe cases of unstable angina. In some centers rates as high as 15 mg/h are used in severe cases of unstable angina or impending myocardial infarction.

If patients treated with ISDN intravenously are to be placed on oral ISDN the rate of infusion should be tapered-off gradually and oral treatment should start at least 1 h before discontinuation of intravenous therapy. Usually oral treatment is initiated 12–24 h before discontinuation of the infusion.

27.2.5. Transcutaneous

ISDN ointment is available, but it has rarely been used clinically since the introduction of transdermal nitroglycerin discs. No transdermal discs of ISDN are available.

27.2.6. Intracoronary

At present it is preferred to use nitroglycerin when intracoronary administration of nitrates is indicated. No dosage recommendations for intracoronary ISDN are available. A new ISDN intracoronary preparation is now evaluated.

27.2.7. Dose Adjustment

No dose adjustments of ISDN are required in the elderly or in patients with renal disease. The dose should be reduced in patients with hepatic dysfunction. No dosage recommendations for children are available.

27.3. IS-5-MN

IS-5-MN may be given by all routes used for other nitrates. However, there is little advantage in giving it sublingually or transdermally. Most studies have therefore been performed with intravenous or oral administration.

27.3.1. Intravenous

Infusion rates of 1–5 mg/h have been used in several studies.

27.3.2. Oral

Recent dose-finding studies have revealed that the optimal oral dose of IS-5-MN is 20 mg. Tauchert et al [428] reported a study on the effect of 5, 10, 20 and 50 mg IS-5-MN on heart rate, systemic and pulmonary arterial pressure, work capacity and duration of work in patients with coronary artery disease. The 5-mg dose had only slightly reduced the pulmonary arterial pressure. The 10-mg dose reduced it by 20% at rest and 22% during exercise. The 20-mg dose produced a near-maximal effect on pulmonary arterial pressure: 25% reduction at rest and 30% during exercise. The 50-mg dose augmented the effect only slightly: by an additional 4% at rest and 10% during exercise. Both the 20-mg and 50-mg doses had similar effects on work capacity and duration of exercise. Therefore the optimal oral dose of IS-5-MN is 20 mg. A higher dose may result in vascular tolerance which will eliminate the additional effect.

Hemodynamic studies by Jansen et al [429] confirmed that a 20-mg dose produces a near-maximal therapeutic effect. The parameter measured in this dose-finding study was reduction in mean pulmonary artery pressure.

The therapeutic oral regimen of conventional tablets of IS-5-MN is 20 mg two to three times daily. In severe cases 40–50 mg three times daily may be used. The same doses are used also in combined therapy.

Akhras and Jacksen [292] reported that while no significant difference in the antianginal effect was observed between doses of 20 and 40 mg, twice daily, individual variations occurred, indicating a need for dosage flexibility. The dosage regimen of IS-5-MN does not require changes in patients with impairment of renal or hepatic function.

E. Conclusions

The marked development in formulations and delivery forms of nitrates has made these agents the most flexible therapeutic class in cardiovascular therapy. For each patient an individually optimalized mode of administration may be tailored. In no case is replacement of a nitrate by a drug from another group required because of inadequate delivery systems.

In the most important indications, namely, angina pectoris and congestive heart failure, nitrates are as effective as or more effective than all other forms of pharmacological treatment. In angina pectoris they form the first line of therapy, both for treatment of the acute event and for prevention of recurrences. Nitrates may be combined with all other forms of antianginal therapy, and graduated antianginal therapy, including nitrates as the first step, is now accepted worldwide as the routine clinical practice.

As the risks of diuretics in congestive heart failure are apprehended (particularly for elderly patients), vasodilators, and primarily nitrates, become the cornerstone of therapy. Graduated therapy of heart failure traditionally includes nitrates as the third step but now also as the second step, immediately after diuretics. Preliminary studies have revealed that some patients with heart failure may be adequately managed with nitrate monotherapy.

An important new field is the use of nitrates together with revascularization procedures. Nitrates can increase the efficacy of thrombolysis and prevent some of the complications of angioplasty.

Two recent reviews by Wenger et al [430] and Dalen et al [431] emphasize the use of nitrates, beta-blockers and antiarrhythmic agents during hospitalization of patients with uncomplicated acute myocardial infarction. In the long-term management of patients after an acute myocardial infarction nitrates are most commonly used. Maseri [432] has stated that, as the positive effect of nitrates on reduction of infarct size, infarct extension and recurrent ischemia is extremely likely, nitrates should be used in patients with acute myocardial infarction from the very moment they reach the hospital, unless contraindicated by hypotension. Thus, a century after their introduction nitrates stand as one of the most important therapeutic tools in cardiovascular medicine and offer more and more exciting fields for future research.

References

1. Needleman P, Lang S, Johnson EM (1972) Organic nitrates: Relationship between biotransformation and rational angina pectoric therapy. J Pharmacol Exp Ther 181:489–497
2. Needleman P, Jakschik B, Johnson EM (1973) Sulfhydryl requirement for relaxation of vascular smooth muscle. J Pharmacol Exp Ther 187:324–331
3. Needleman P, Johnson EM (1973) Mechanism of tolerance development to organic nitrates. J Pharmacol Exp Ther 184:709–715
4. Ignarro LJ, Gruetter CA (1980) Requirement of thiols for activation of coronary arterial guanylate cyclase by glyceryl trinitrate and sodium nitrite: Possible involvement of 5-nitrosothiols. Biochim Biophys Acta 631:221–231
5. Horowitz JD, Antman EA, Lorell BH et al (1983) Potentiation of the cardiovascular effects of nitroglycerin by N-acetylcysteine. Circulation 68:1247–1253
6. Levin RI, Jaffe EA, Weksler BB et al (1981) Nitroglycerin stimulates synthesis of prostacyclin by cultured human endothelial cells. J Clin Invest 67:762–769
7. Morcillio E, Reid PR, Dubin N et al (1980) Myocardial prostaglandin E release by nitroglycerin and modification by indomethacin. Am J Cardiol 45:53–57
8. McGregor M, Crelinsten GL (1984) Possible interactions of indomethacin and nitrates in angina pectoris. Am J Cardiol 53:225–227
9. Feigen LP, Chapnick BM, Flemming JE et al (1978) Prostaglandins: renal vascular responses to bradykinin, histamine and nitroglycerin. Am J Physiol 234:H496–H502
10. Förster W (1980) Significance of prostaglandins and thromboxane. A_2 for the mode of action of carciovascular drugs. In: Samuelsson B, Ramwell PW, Paoletti R (eds) Advances in prostaglandin and thromboxane research, vol 7. Raven, New York, 609–618

10a. Graven PA, DeRuhertis FR (1978) Restoration of the responsiveness of purified guanylate cyclase to nitrosoguanidine, nitric oxide, and related activators by heme and heme proteins: evidence for the involvement of the paramagnetic nitrosyl-heme complex in enzyme activation. J Biol Chem 253:8433–8443

10b. Furchgott RE, Zawadzki JV (1980) The obligatory role of endothelial cells in the relaxation of arterial smooth muscle by acetylcholine. Nature (London) 188:373–376

10c. Ignarro LJ, Kadowitz PJ (1985) The pharmacological and physiological role of cyclic GMP in vascular smooth muscle relaxation. Annu Rev Pharmacol Toxiol 25:171–191

10d. Ignarro LJ, Burke TM, Wood KS, Wolin MS, Kadowitz PJ (1984) Association between cyclic GMP accumulation and acetylcholine-elicited relaxation of bovine intrapulmonary artery. J Pharmacol Exp Ther 228:682–690

10e. Ignarro IJ, Buga GM, Woon KS, Byrns RE, Chaudhuri G (1987) Endothelium-derived relaxing factor produced and released from artery and vein is nitric oxide. Proc Natl Arad Sci USA 84:9265–9269

10f. Ignarro IJ, Byrns RE, Boga GM, Wood KS (1987) Endotheliumderived relaxing factor from pulmonary artery and vein possesses pharmacological and chemical properties that are identical to those for nitric oxide radical. Circ Res 61:868–879

10g. Palmer RMJ, Ferrige AG, Moncada S (1987) Nitric oxide release accounts for the biological activity of endothelium-derived relaxing factor. Nature (London) 327:624–626

11. Danahy DT, Burwell DT, Aronow WS, Prakash R (1977) Sustained hemodynamic and antianginal effects of high oral dose isosorbide dinitrate. Circulation 55:381–387

12. Lee G, Mason DT, Amsterdam EA, De Maria AN (1978) Antianginal efficacy of oral therapy with isosorbide dinitrate capsules: prolonged benefit shown by exercise testing in patients with ischemic heart disease. Chest 73:327–332
13. Thadani U, Fung H, Darke AC, Parker JO (1980) Oral isosorbide dinitrate in angina pectoris: Comparison of duration of action and dose-response relation during acute and sustained therapy. Am J Cardiol 49:411–419
14. Strauch M, Grewe N (1979) Die Wirkung von isosorbiddinitrat, Isosorbid-2 und 54-Mononitrat auf das Belastungs-EKG und auf die Hämodynamik während Vorhofstimulation bei Patienten mit Angina Pectoris. Z Kardiol 68:687–693
15. Feldman RL, Pepine CJ, Curry RC, Conti R (1979) Coronary arterial responses to graded doses of nitroglycerin. Am J Cardiol 43:91–97
16. Brown BG, Bolson E, Petersen RB, Pierce CD, Dodge HT (1981) The mechanisms of nitroglycerin action: stenosis vasodilation as a major component of the drug response. Circulation 64:1089–1097
17. Schnaar RL, Sparks HV (1972) Response of large and small coronary arteries to nitroglycerin, $NaNO_2$, and adenosine. Am J Physiol 223:223–228
18. Harder DR, Belardinelli L, Sperelakis N, Rubio R, Berne RM (1979) Differential effects of adenosine and nitroglycerin on the action potentials of large and small coronary arteries. Circ Res 44:176–182
19. Simon R, Koch M, Reil G-H et al (1987) Effect of intracoronary nitroglycerin on coronary vascular volume in man assessed by a double indicator method. Eur Heart J 8 (Suppl 1):22
19a. Simonetti I, Winniford MD, Rossen JD, Marcus ML (1988) Biphasic effect of nitroglycerin on the human coronary circulation. Eur Heart J 9 (Suppl 1):329 (1826) abstr
20. Klein RC, Grehl TM, Stengert KB, Mason DT (1981) Evaluation of the effects of systemic nitroglycerin on perfusion of ischemic myocardium in coronary heart disease assessed intraoperatively by antegrade blood flow through intact saphenous vein bypass grafts. Am Heart J 101:292–299
21. Benchimol A, Stegall HF, Gartlan JL (1971) New method to measure phasic coronary blood velocity in man. Am Heart J 81:93–101
22. Simon R, Amende I, Lichtlen PR (1981) Effects of nitroglycerin on blood velocity and flow in coronary arteries and bypass grafts in man. In: Nitrates III. Lichtlen PR, Engel H-J, Schrey A, Swan HJC (eds) Springer-Verlag, Berlin 202–208
23. Fujita M, McKown DP, McKown MD, Franklin D (1986) Comparative effects of nitroglycerin and diltiazem on regressed, newly developed coronary collaterals in conscious dogs. J Am Coll Cardiol 7 (Suppl):53A
24. Tono-oka I, Satoh S, Hoshi H, Tsuiki K, Yasui S (1986) Actual enhancement of myocardial perfusion during exercise after isosorbide dinitrate (ISDN) infusion. X World Congress of Cardiology, Washington, D.C., September 14–19 (2956)
24a. Okazaki Y, Mishima M, Sato H, Nanto S, Komamura K, Kodama K (1989) Effect of nitroglycerin on coronary collateral function during exercise. Cardiovasc Drugs & Ther 3 (Suppl 2):619 abstr
25. Feldman RL, Pepine CJ, Curry RC, Conti CR (1978) Case against routine use of glyceryl trinitrate before coronary angiography. Br Heart J 40:992–997
26. Melchior JP, Doriot P-A, Urban P, Noble J, Rutishauser W (1987) Nitrates and coronary stenoses: Relative and absolute measurements by densitometry. Eur Heart J 8 (Suppl 1):21
27. Hombach V, Jansen WC, Behrenbeck DW, Tauchert M, Niehues B, Hilger H (1981) Influence of nitroglycerin on myocardial metabolism of cyclic AMP, cyclic GMP, lactate, free fatty acids, and glucose at rest and during exercise in patients with coronary heart disease. In: Nitrates III. Lichtlen PR, Engel H-J, Schrey A, Swan HJC (eds) Springer-Verlag, Berlin 401–405
28. Ganz W, Marcus HS (1972) Failure of intracoronary nitroglycerin to alleviate pacing-induced angina. Circulation 46:880–889
29. DePace NL, Herling IM, Kotler MN, Hakki AH, Spielman SR, Segal BL (1982) Intravenous nitroglycerin for rest angina. Arch Intern Med 142:1806–1809

30. Krayenbühl HP, Hess OM, Gage JE, Grimm J (1987) Nitroglycerin prevents coronary artery stenosis narrowing during dynamic exercise. Eur Heart J 8 (Suppl 1):23

30a. Nonogi H, Hess QM, Ritter M, Bortone A, Corin WJ, Grimm J, Krayenbuehl HP (1988) Prevention of coronary vasoconstriction by diltiazem during dynamic exercise in patients with coronary artery disease. Am Coll Cardiol 12:892–899

31. Goldstein RE, Bennett ED, Leech GL (1979) Effect of glyceryl trinitrate on echocardiographic left ventricular dimensions during exercise in the upright position. Br Heart J 42:245–254

32. Fuchs RM, Brinker JA, Guzman PA, Kross DE, Yin FCP (1983) Regional coronary blood flow during relief of pacing-induced angina by nitroglycerin. Am J Cardiol 51:19–23

33. Engel HJ, Wolf R, Pretschner P, Hundeshagen H, Lichtlen PR (1981) Effect of nitrates on myocardial blood flow during angina: Comparison of results obtained by inert gas clearance and 201 thallium imaging. In: Nitrates III. Lichtlen PR, Engel H-J, Schrey A, Swan HJC (eds) Springer-Verlag, Berlin 184–191

34. McAnulty JH, Hattenhauer MT, Rosch J, Kloster FE, Rahimtoola SH (1975) Improvement in left ventricular wall motion following nitroglycerin. Circulation 51:140–145

35. Reddy SP, Curtis EI, O'Toole JD et al (1975) Reversibility of left ventricular asynergy by nitroglycerin in coronary artery disease. Am Heart J 90:479–486

36. Dumesnil JG, Ritman EL, Davis GD, Gau GT, Rutherford BD, Frye RL (1975) Regional left ventricular wall dynamics before and after sublingual administration of nitroglycerin. Am J Cardiol 36:419–425

37. Bodenheimer MM, Banka VS, Hermann GA, Trout RG, Pasdar H, Helfant RH (1976) Reversible asynergy, histopathologic and electrographic correlations in patients with coronary artery disease. Circulation 53:792–796

38. Helfant RH, Pine R, Meister SG, Feldman MS, Trout RG, Banka VS (1974) Nitroglycerin to unmask reversible asynergy. Correlation with post coronary bypass ventriculography. Circulation 50:108–113

39. Chesebro JH, Ritman EL, Frye RL et al (1976) Regional myocardial wall thickening response to nitroglycerin – a predictor of myocardial response to aortocoronary bypass surgery. Circulation 57:952–957

40. Pepine CJ, Feldman RL, Ludbrock P et al (1986) Left ventricular dyskinesia reversed by intravenous nitroglycerin: a manifestation of silent myocardial ischemia. Am J Cardiol 58:38B–42B

41. Karsch KR, Rentrop P, Neuhaus KL, Bornikoel K, Kreuzer H (1980) Right ventricular haemodynamics before and after nitroglycerin in patients with coronary heart disease. In: Nitrates II. Rudolph W, Schrey A (eds) Urban & Schwarzenberg Munich 418–428

41a. Plotnick GD, Kahn B, Rogers WJ, Fisher ML, Becker LC (1988) Effect of postural changes, nitroglycerin and verapamil on diastolic ventricular function as determined by radionuclide angiography in normal subjects. J Am Coll Cardiol 12:121–129

41b. Unterberg R, Gruber H, Vökjer W, Karsch K (1989) Right ventricular diastolic function during acute alterations in loading conditions. Eur Heart J 10 (Suppl):349 (P1973) abstr

42. Ludbrook PR, Byrne JD, Kurnik PB, McKnight RC (1977) Influence of reduction of preload and afterload by nitroglycerin on left ventricular diastolic pressure-volume relation and relaxation in man. Circulation 56:937–943

43. Amende I, Simon R, Hood WP Jr, Lichtlen PR (1983) Effects of nitroglycerin on left ventricular diastolic properties in man. Z Kardiol 72 (Suppl 3):62–65

44. Hirzel HO, Stoffel P, Krayenbuehl HP (1983) Does Isosorbide-5-Mononitrate influence left ventricular relaxation? Z Kardiol 72 (Suppl 3):66–70

44a. Bonow RO, Bacharach SL, Green MV, Kent KM, Rosing DR, Lipson LC, Leon MB, Epstein SE (1981) Impaired left ventricular diastolic filling in patients with coronary artery disease: assessment with radio-nuclide cineangiography. Circulation 64:315–323

44b. Fouad FM (1987) Left ventricular diastolic function in hypertensive patients. Circulation 75 (Suppl I):48–55

44c. Inouye I, Massie B, Loge D, Topic N, Silverstein D, Simpson P, Tubau J (1984) Abnormal left ventricular filling: an early finding in mild to moderate systemic hypertension. Am J Cardiol 53:120–126
44d. Yamagishi T, Uki K, Yamauchi M, Yamada H, Kohno M, Kumada T, Ozaki M, Kusukawa R (1986) Acute effects of sublingual isosorbide dinitrate on global and regional left ventricular diastolic filling in normal persons. Am J Cardiol 58:1061–1066
45. De Caterina R, Gianessi D, Crea F, Chierchia S, Bernini W, Gazzetti P, L'Abbate A (1983) Antiplatelet effects of isosorbide dinitrate in man. Z Kardiol 72 (Suppl 3):46–51
46. Schrör K, Grodzinska L, Darius H (1981) Stimulation of coronary vascular prostacyclin and inhibition of human platelet thromboxane A_2 after low-dose nitroglycerin. Thromb Res 23:59–67
47. Rolland PH, Berenger EP, Cano JP (1987) The anti-thrombogenic endothelial cell defense (ATECD). ATECD-based characteristics in cultured endothelial cells (EC) and ATECD modulation by short- and long-term exposure of EC to isosorbide nitrates (ISN). Eur Heart J 8 (Suppl 1):4
47a. Stamler J, Cunningham M, Loscalzo J (1988) Reduced thiols and the effect of intravenous nitroglycerin on platelet aggregation. Am J Cardiol 62:377–380
47b. Brügger W, Imhof P, Müller P, Moser P, Reubi F (1985) Effect of nitroglycerin on blood rheology in healthy subjects. Eur J Clin Pharmacol 29:331–336
48. Hopf R, Schmidt H, Kaltenbach M (1978) Kombination von Isosorbid-dinitrat mit Nifedipin bei der Behandlung der koronaren Herzkrankheit. Z Kardiol (Suppl 5):67
49. De Ponti C, Mauri F, Ciliberto GR, Carù B (1979) Comparative effects of nifedipine, verapamil, isosorbide dinitrate and propranolol on exercise-induced angina pectoris. Eur J Cardiol 10:47–58
50. Bassan MM, Weiler-Ravell D, Shaley O (1983) Comparison of the antianginal effectiveness of nifedipine, verapamil, and isosorbide dinitrate in patients receiving propranolol: a double-blind study. Circulation 68:568–575
51. Schulz W, Tatzel R, Jost S, Kaltenbach M, Kober G (1985) Clinical importance of the coronary steal phenomenon: Are the coronary steal phenomenon after dipyridamole and the antianginal efficacy of isosorbide dinitrate or nifedipine related to coronary anatomy? In: Mononitrates. Cohn JN, Rittinghausen R (eds) Springer-Verlag Berlin 254–264
51a. Engby B, Dehlholm B, Bötker H-E, Rickers H, Thygesen K (1989) Stable angina treated with isosorbide-5-mononitrate, durules®, diltiazem or the combination. Eur Heart J 10 (Suppl):72 (350) abstr
52. Packer M, Lee WH, Medina N, Yushak M (1986) Should nitrates or nifedipine be preferred as first-line anti-anginal therapy in patients with left ventricular dysfunction. J Am Coll Cardiol 7 (Suppl):178A
53. Stegaru B, Loose R, Keller H, Buss J, Wetzel E (1989) Effects of long-term treatment with 120 mg sustained-release ISDN and 60 mg sustained-release nifedipine on myocardial perfusion. Am J Cardiol (Suppl) 61 (9):74E–77E
54. Severi S, Davies G, Maseri A, Marzullo P, L'Abbate A (1980) Long-term prognosis of "variant" angina with medical treatment. Am J Cardiol 46:226–232
55. Raizner AE, Chahine RA, Gaston WR, Smith DL, Luchi RJ, Miller RR (1981) The effectiveness of combined verapamil and nitrate therapy in Prinzmetal's variant angina. In: Calcium Antagonism in Cardiovascular Therapy: Experience with Verapamil. Zanchetti A, Krikler DM (eds) Amsterdam, Excerpta Medica, 191–197
56. Lablanche JM, Beuscart R, Traisnel G, Fourrier JL, Bertrand ME (1986) Additive effects of nitrates and calcium antagonists. J Am Coll Cardiol 7 (Suppl):28A
57. Abrams J, Hoekenga D (1986) Failure of combination antianginal therapy with diltiazem and nitroglycerin: Are two drugs better than one? J Am Coll Cardiol 7 (Suppl):27A
57a. Apollo WP, Zeils R, Gascho JA (1989) Nifedipine does not alter the venodilatory response to nitroglycerin. J Am Coll Cardiol 13 (2):154A abstr
58. Tolins M, Weir EK, Chester E, Pierpont GL (1983) "Maximal" drug therapy is not necessarily optimal in chronic stable angina. Circulation 68 (Suppl III):209

59. Katz RI, Weintraub WS, Bodenheimer MM, Agarwal JB, Banka VS, Helfant RH (1982) Failure of low dose nifedipine to improve exercise tolerance in stable angina. Am J Cardiol 49:895
60. Silber S, Vogler A, Theisen K (1986) Problems of anti-ischemic drug therapy in patients with insufficient nitrate response. J Am Coll Cardiol 7 (Suppl):28A
61. Collins P, Sheridan D (1985) Improvement in angina pectoris with alpha adrenoreceptor blockade. Br Heart J 53:488–492
62. Rafflenbeul W, Lichtlen PR (1983) Release of residual vascular tone in coronary artery stenoses with nifedipine and glyceryl trinitrate. In: 5th International Adalat Symposium. New Therapy of Ischemic Heart Disease and Hypertension, Kaltenbach M, Neufeld HN (eds) Excerpta Medica, Amsterdam, 300–317
63. Ohmori Y, Kajiya T, Nakashima Y, Shiotani H, Maedy K, Fukuzaki H (1983) Comparative effects of intravenous nitroglycerin and diltiazem given during exercise-induced ischemia in patients with effort angina. Circulation 68 (Suppl III):207
63a. Choong CYP, Freedman B, Roubin GS, Shen WF, Bautovich GJ, Harris PH, Kelly DT (1989) Comparison of the effects of nifedipine and nitroglycerin on hemodynamic determinants of myocardial oxygen consumption and supply during exertional angina. J Cardiovasc Pharmacol 13 (3):361–369
64. Hill JA, Feldman RL, Pepine CJ, Conti CR (1982) Randomized double-blind comparison of nifedipine and isosorbide dinitrate in patients with coronary arterial spasm. Am J Cardiol 49:431–438
65. Ginsburg R, Lamb IH, Schroeder JS, Ju M, Harrison DC (1982) Randomized double-blind comparison of nifedipine and isosorbide dinitrate therapy in variant angina pectoris due to coronary artery spasm. Am Heart J 103:44–48
66. Muller JE, Turi ZG, Pearle DL et al (1984) Nifedipine and conventional therapy for unstable angina pectoris: a randomized, double-blind comparison. Circulation 69:728–739
67. Schneider WU, Bussmann WD, Stahl B, Kaltenbach M (1984) Dose-response relation of antianginal activity of isosorbide dinitrate. Am J Cardiol 53:700–705
68. Julian DG (1985) Comparison and combinations in anti-anginal therapy. Eur Heart J 6 (Suppl):37–45
69. Deanfield J, Wright C, Fox K (1983) Treatment of angina pectoris with nifedipine: importance of dose titration. Br Med J 286:1467–1470
70. Betablocker Heart Attack Trial Research Group (1983) A randomized trial of propanolol in patients with acute myocardial infarction. II. Morbidity results. JAMA 250:2814–2819
70a. Wallace WA, Wellington KL, Murphy GW, Liang C-S (1989) Comparison of antianginal efficacies and exercise hemodynamic effects of nifedipine and diltiazem in stable angina pectoris. Am J Cardiol 63:414–418
70b. Angelin PF (1983) Hemodynamic and enzymatic effects of intravenous nifedipine administration in acute myocardial infarction. 5th International Adalat Symposium. In: New Therapy of Ischemic Heart Disease and Hypertension. Kaltenbach M, Neufeld HN (eds) Excerpta Medica, Amsterdam, pp 208–214
70c. Fox KM (1983) Treatment of chronic stable angina pectoris with nifedipine. 5th International Adalat Symposium. In: New Therapy of Ischemic Heart Disease and Hypertension. Kaltenbach M, Neufeld HN (eds) Excerpta Medica, Amsterdam, pp 197–208
70d. Subramanian VB (1982) Rationale for the choice of calcium antagonists in chronic stable angina. An objective double-blind placebo-controlled comparison of nifedipine and verapamil. Am J Cardiol 50:1173–1178
70e. The Israeli Sprint Study Group (1988) Secondary prevention reinfarction Israeli nifedipine trial (SPRINT). A randomized intervention trial of nifedipine in patients with acute myocardial infarction. Eur Heart J 9:354–364
70f. The Sprint Group (1988) The secondary prevention reinfarction Israeli nifedipine trial (Sprint II). Eur Heart J 9:350 abstr
70g. Frishman W, Charlap S, Kimmel B, Teicher M, Cinnamon J, Allen L, Strom J (1988) Diltiazem, nifedipine, and their combination in patients with stable angina pectoris:

effects on angina, exercise tolerance, and the ambulatory electrocardiographic ST segment. Circulation 77 (4):774–786
70h. Smith HJ, Goldstein RA, Griffith JM, Epstein SE (1976) Regional contractility: selective depression of ischemic myocardium by verapamil. Circulation 54:629–635
70i. Friedman MJ, Temkin LP, Goldman S, Ovitt TW (1983) Effects of propranolol on resting and post extra systolic potentiated left ventricular function in patients with coronary artery disease. Am Heart J 106:81–84
70j. Port S, Gobb FR, Jones RH (1980) Effects of propranolol on left ventricular function in normal men. Circulation 61 (2):358–365
70k. Milne JR, Camm AJ, Ward DE, Spurrell RAJ (1980) Effect of intravenous propranolol on QT interval: a new method of assessment. Br Heart J 43:1–7
70l. Norwegian Multicenter Study Group (1981) Timolol-induced reduction in mortality and reinfarction in patients surviving acute myocardial infarction. N Engl J Med 304:801–807
70m. Beta-Blocker Heart Attack Trial Research Group (1982) A randomized trial of propranolol in patients with acute myocardial infarction: I. Mortality results. JAMA 247:1707–1714
70n. Rehnqvist N, Olsson G (1983) Influence on ventricular arrhythmias by chronic postinfarction treatment with metoprolol. Circulation 68 (Suppl 3):369 abstr 1476
70o. The MIAMI Trial Research Group (1985) Metoprolol in acute myocardial infarction (MIAMI). A randomized placebo-controlled international trial. Eur Heart J 6:199–226
70p. Ertl G, Kloner RA, Alexander RW, Braunwald E (1982) Limitation of experimental infarct size by an angiotensin-converting enzyme inhibitor. Circulation 65:40–48
70q. Van Gilst WH, De Graeff PA, Wesseling H, De Langen CDJ (1986) Reduction of reperfusion arrhythmias in the ischemic isolated rat heart by angiotensin converting enzyme inhibitors: A comparison of captopril, enalapril, and HOE 498. J Cardiovasc Pharmacol 8:722–728
70r. Van Gilst WH, De Graeff PA, Kingma JH, Wesseling H, De Langen CDJ (1984) Captopril reduces purine loss and reperfusion arrhythmias in the rat heart after coronary artery occlusion. Eur J Pharmacol 100:113–117
70s. Linz W, Schölkens BA, Han Y-F (1986) Beneficial effects of the converting enzyme inhibitor, ramipril, in ischemic rat hearts. J Cardiovasc Pharmacol 8 (Suppl 10):91–99
70t. Pfeffer JM, Pfeffer MA, Braunwald E (1985) Influence of chronic captopril therapy on the infarcted left ventricle of the rat. Circ Res 57:84–95
70u. Gibbs JSR, Crean PA, Mockus L, Sutton GC, Fox KM (1989) The variable effects of angiotensin converting enzyme inhibition on myocardial ischaemia in chronic stable angina. Eur Heart J 10 (Suppl):V abstr 417
70v. Brown BG, Bolson E, Petersen RB, Pierce DC, Dodge HT (1981) The mechanisms of nitroglycerin action: stenosis vasodilation as a major component of the drug response. Circulation 64:1089–1097
70w. Goldstein RE, Bennett ED, Leech GL (1979) Effect of glyceryl trinitrate on echocardiographic left ventricular dimensions during exercise in the upright position. Br Heart J 42:245–254
70x. Packer M (1989) Combined beta-adrenergic and calcium-entry blockade in angina pectoris. N Engl J Med 320 (11):709–718
70y. Peycelon P, Lusson JR, Abrieu V, Veyre A, Cassagnes J (1989) T1201 tomoscintigraphy during isosorbide dinitrate infusion (ISDN-SPECT): A predictive test for recoverable myocardium. Eur Heart J 10 (Suppl):57 (288) abstr
71. Distante A, Sabino F, L'Abbate A (1987) Cooling down unstable angina with high dosage of isosorbide dinitrate continuously infused. Eur Heart J 8 (Suppl 1):36
72. Page A, Gateau P, Ohayon J, Coupillaud J, Le Minh D, Besse P (1981) Intravenous nitroglycerin in unstable angina. In: Nitrates III. Lichtlen PR, Engel H-J, Schrey A, Swan HJC (eds) Springer-Verlag, Berlin 371–376
73. Rafflenbeul W, Bossaller C, Lichtlen Pr (1986) Intravenous nifedipine versus nitroglycerin in patients with angina at rest. J Am Coll Cardiol 7 (Suppl):179A
74. Horowitz JD, Henry CA, Syrjanen ML et al (1987) Nitroglycerin/N-acetylcysteine in the management of unstable angina pectoris. Eur Heart J 8 (Suppl 1):24

74a. Flaherty JT (1989) Nitrate tolerance: A review of the evidence. Drugs 37 (4):523–550
75. Araujo LI, Camici P, Spinks TJ, Jones t, Maseri A (1987) Myocardial metabolic changes associated to administration of nitrates in patients with unstable angina as assessed by positron emission tomography. Eur Heart J 8 (Suppl. 1):26
76. Prinzmetal M, Kennamer R, Merliss RR, Wada T, Bor N (1959) Angina pectoris. I. A variant form of angina pectoris. Am J Med 27:375–388
77. Guazzi M, Polese A, Fiorentini G, Magrini F, Bartorelli C (1971) Left ventricular performance and related hemodynamic changes in Prinzmetal's variant angina pectoris. Br Heart J 33:89–94
78. Nelson C, Nowak B, Childs H, Weinrauch L, Forwand S (1977) Provocative testing for coronary arterial spasm: rationale risk and clinical illustrations. Am J Cardiol 40:624–629
79. Baumann D (1978) Complications after provocation of coronary spasm with ergonovine maleate. Am J Cardiol 42:694–695
80. Schroeder JS, Bolen JL, Quint A et al (1977) Provocation of coronary spasm with ergonovine maleate. Am J Cardiol 40:487–491
81. Epstein SE, Goldstein RE, Redwood DR, Kent KM, Smith ER (1973) The early phase of acute myocardial infarction: pharmacologic aspects of therapy. Ann Intern Med 78:918–936
82. Buxton A, Goldberg S, Hirshfeld JW et al (1980) Refractory ergonovine-induced coronary vasospasm: importance of intracoronary nitroglycerin. Am J Cardiol 46:329–334
83. Pepine CJ, Feldman RL, Conti CR et al (1982) Action of intracoronary nitroglycerin in refractory coronary artery spasm. Circulation 65:411–414
84. Curfman GD, Heinsimer JA, Lozner EC, Fung H-L (1983) Intravenous nitroglycerin in the treatment of spontaneous angina pectoris: a prospective, randomized trial. Circulation 67:276–282
85. Kaplan K, Davison R, Parker M, Przybylek J, Teagarden JR, Lesch M (1983) Intravenous nitroglycerin for the treatment of angina at rest unresponsive to standard nitrate therapy. Am J Cardiol 51:694–698
86. Distante A, Maseri A, Severi S, Biagini A, Chierchia S (1979) Management of vasospastic angina at rest with continuous infusion of isosorbide dinitrate. Am J Cardiol 44:533–539
87. Winniford MD, Gabliani G, Johnson SM, Mauritson DR, Fulton KL, Hillis LD (1984) Concomitant calcium antagonist plus isosorbide dinitrate therapy for markedly active variant angina. Am Heart J 108:1269–1273
88. Bethge HD, Bachmann K (1983) Assessment of sublingual isosorbide dinitrate and intracoronary nifedipine during ergonovine provocation. Z Kardiol 72 (Suppl 3):92–97
89. Distante A, Lombardi M, Moscarelli E et al (1985) Efficacy of isosorbide-5-mononitrate in the prevention of induced attacks of vasospastic angina: preliminary results. In: Mononitrates. Cohn JN, Rittinghausen R (eds) Springer-Verlag, Berlin 235–244
90. Franciosa JA, Cohn JN (1980) Sustained hemodynamic effects without tolerance during long-term isosorbide dinitrate treatment of chronic left ventricular failure. Am J Cardiol 45:648–654
91. Zelis R, Mason DT (1975) Isosorbide effect on the vasodilator response to nitroglycerin. JAMA 234:166–170
92. Leier CV, Huss P, Magorien RD, Unverferth DV (1983) Improved exercise capacity and differing arterial and venous tolerance during chronic isosorbide dinitrate therapy for congestive heart failure. Circulation 67:817–822
93. Parker JO, Fung H-L, Ruggirello BS, Stone JA (1983) Tolerance to isosorbide dinitrate: rate of development and reversal. Circulation 68:1074–1080
94. Sutton SC, Fung H-L (1983) Effect of dosage regimen on the development of tolerance to nitroglycerin in rats. J Cardiovasc Pharmacol 5:1086–1092
95. Stewart DJ, Elsner D, Sommer O, Holtz J, Bassenge E (1986) Altered spectrum of nitroglycerin action in long-term treatment: nitroglycerin-specific venous tolerance with maintenance of arterial vasodepressor potency. Circulaton 74:573–582
96. Manyari DE, Smith ER, Spragg J (1985) Isosorbide dinitrate and glyceryl trinitate: demonstration of cross tolerance in the capacitance vessels. Am J Cardiol 55:927–931

97. Schneider W, Kraus M, Zeidler M et al (1987) Dose-dependent relaxation of human venous vessel strips with regard to chronic nitrate pretreatment. Eur Heart J 8 (Suppl 1):6
98. Bennett D, Barclay SA, Adam J, Valentine H, Boyle GJ (1987) Transdermal nitroglycerin causes plasma volume expansion which may explain tolerance to transdermal nitroglycerin. Eur Heart J 8 (Suppl 1):18
99. Wortmann A, Bachmann K, Deuber HJ, Roth E (1986) Hämodynamik und Koronardynamik unter ISDN-Langzeittherapie. Z Kardiol 75 (Suppl 3):68–76
100. Distante A, Lombardi M, Morales MA, Moscarelli E, L'Abbate A (1984) Preserved hemodynamic response to acute dose of ISDN in anginal patients chronically treated with the same drug percutaneously. Eur Heart J 5 (Abstr Suppl 1):269
101. Thadani U, Manyari D, Parker JO, Fung H-L (1980) Tolerance to the circulatory effects of oral isosorbide dinitrate. Rate of development and cross-tolerante to glyceryl trinitrate. Circulation 61:526–535
102. Davidov ME, Mroczek WJ (1977) The effect of sustained-release nitroglycerin capsules on anginal frequency and exercise capacity. Angiology 28:181–189
103. Reichek N, Goldstein RE, Redwood DR, Epstein SE (1974) Sustained effects of nitroglycerin ointment in patients with angina pectoris. Circulation 50:348–352
104. Winsor W, Berger HJ (1975) Oral nitroglycerin as a prophylactic antianginal drug: clinical, physiologic and statistical evidence of efficacy based on a threephase experimental design. Am Heart J 90:611–626
105. Niederer WE, Bethge HD, Bachmann K (1983) Hemodynamic and ventricular dynamic investigations of nitrate tolerance. In: Kaltenbach M, Kober G (eds) Nitrates and Nitrate Tolerance in Angina Pectoris, Steinkopff, Darmstadt, 65–74
106. Schneider W, Wietschoreck A, Bussmann W-D, Kaltenbach M (1983) Sustained antianginal efficacy of oral high-dose isosorbide dinitrate in patients with coronary heart disease. Z Kardiol 72 (Suppl 3):259–267
107. Schwarzer C, Mlczoch J (1982) Development of tolerance under chronic therapy with isosorbide-5-mononitrate. Herz/Kreisl 14:585–589
108. Dalal JJ, Yao L, Parker JO (1983) Nitrate tolerance: influence of isosorbide dinitrate on the hemodynamic and antianginal effects of nitroglycerin. J Am Coll Cardiol 2:115–120
109. Nyberg G, Nilsson B (1985) Antianginal efficacy of isosorbide-5-mononitrate in a controlled-release (Durules) formulation. Br J Chlin Pharmac 20:540P
109a. Unverdorben M, Pfeiffer F, Kunkel B, Bachmann K (1989) Hemodynamic tolerance during a 24h-infusion of ISDN? Cardiovasc Drugs & Ther 3 (Suppl 2):637 abstr
110. Ohlmeier H, Mertens HM, Mannebach H, Gleichmann U (1985) Tolerance and rebound phenomena in nitrate therapy. In: Mononitrates. Cohn JN, Rittinghausen R (eds) Springer-Verlag, Berlin 107–123
111. Rennhak U, Riebesel, T, Biamino G (1985) A double-blind cross-over study on the effectiveness and possible development of tolerance during long-term therapy with isosorbide-5-mononitrate or isosorbide dinitrate slow-release in coronary artery disease. In: Mononitrates. Cohn JN, Rittinghausen R (eds) Springer-Verlag, Berlin 147–153
112. Silber S, Krause KH, Garner G, Theisen K, Jahrmarker H (1983) Anti-ischemic effects of an 80 mg tablet of isosorbide dinitrate in sustained-release form before and after 2 weeks of treatment with 80 mg once daily or twice daily. Z Kardiol 72 (Suppl 3):211–217
113. Jansen W, Osterspey A, Tauchert M (1982) 5-Isosorbide mononitrate at rest and on exercise in coronary heart disease: acute and long-term effects. Dtsch Med Wschr 107:1499–1506
114. Parker JO, VanKoughnett KA, Fung H-L (1984) Transdermal isosorbide dinitrate in angina pectoris: effect of acute and sustained therapy. Am J Cardiol 54:8–13
115. Parker JO, Fung H-L (1984) Transdermal nitroglycerin in angina pectoris. Am J Cardiol 54:471–476
116. Kaltenbach M, Schneider W (1986) Continuing antianginal action during chronic nitrate treatment despite cessation of haemodynamic effects. Dtsch Med Wschr 111:383–386
116a. Jugdutt BI, Warnica JW (1989) Tolerance with low dose intravenous nitroglycerin therapy in acute myocardial infarction. Am J Cardiol 64:581–587

117. Blasini R, Reiniger G, Brügmann C (1985) Tolerance to the antiischemic effect of isosorbide dinitrate during continuous but not during intermittent oral therapy. In: Mononitrates. Cohn JH, Rittinghausen R (eds) Springer-Verlag, Berlin 124–129
118. Boertz A, Bonn R (1986) Nitrate therapy without loss of action by correct dosage. Z Kardiol 75 (Suppl 3):57–60
118a. Naito H, Matsuda Y, Shiomi K, Yorozu T, Maeda T, Lee H, Seki K, Nakashima H (1989) Effects of sublingual nitrate in patients receiving sustained therapy of isosorbide dinitrate for coronary artery disease. Am J Cardiol 64:565–568
119. Margolis JR, Saive H, Glemser E (1983) Intravenous bolus nitroglycerin – hemodynamic effects and clinical implications. Z Kardiol 72 (Suppl 3):246–250
120. Axelsson KL, Andersson RGG (1983) Tolerance towards nitroglycerin, induced in vivo, is correlated to a reduced cGMP response and an alteration in cGMP turnover. Eur J Pharmacol 88:71–79
121. Kukovetz WR, Holzmann S, Romanin C (1987) Mechanism of vasodilation by nitrates: role of cyclic GMP. Cardiology 74 (Suppl 1):12–19
121a. Münzel T, Holtz J, Mülsch A, Stewart DJ, Bassenge E (1989) Nitrate tolerance in epicardial arteries on in the venous system is not reversed by N-acetylcystein in vivo, but tolerance-independent interactions exist. Circulation 79:188–198
121b. Feng J, Feng X-H, Schneeweiss A (1990) The efficacy of isosorbide-5-mononitrate on painful and silent mypocardial ischemia in patients after myocardial infarction. Am J Cardiol (in press)
121c. Loaldi A, Polese A, Montorsi P, De Cesare N, Fabbiocchi F, Ravagnani P, Guazzi MD (1989) Comparison of nifedipine, propranolol and isosorbide dinitrate on angiographic progression and regression of coronary arterial narrowing in angina pectoris. Am J Cardiol 64:433–439
122. Shimoura K, Meerbaum S, Sakamaki T et al (1983) Relation between functional response to nitroglycerin and extent of myocardial necrosis in dogs; mapping of the left ventricle by 2-dimensional echocardiography. Am J Cardiol 52:177–183
123. Komer RR, Edalji A, Hood WB (1979) Effects of nitroglycerin on echocardiographic measurements of left ventricular wall thickness and regional myocardial performance during acute coronary ischemia. Circulation 59:926–937
124. Delgado CE, Pitt B, Taylor DR, Weisfeldt M, Kelly DT (1975) Role of sublingual nitroglycerin in patients with acute myocardial infarction. Br Heart J 37:392–396
125. Theroux P, Ross J, Franklin D, Kemper WS, Sasayama S (1976) Regional myocardial function in the conscious dog during acute coronary occlusion and responses to morphine, propranolol, nitroglycerin, and lidocaine. Circulation 53:302–314
126. Ramanathan KB, Bodenheimer MM, Banka VS, Helfant RH (1979) Severity of contraction abnormalities after acute myocardial infarction in man: response to nitroglycerin. Circulation 60:1230–1237
127. Theroux P, Franlin D, Ross J Jr et al (1974) Regional myocardial function during acute coronary artery occlusion and its modification by pharmacologic agents in the dog. Circ Res 35:896–908
128. Banka VS, Bodenheimer MM, Helfant RH (1975) Nitroglycerin in experimental myocardial infarction. Effects on regional left ventricular length and tension. Am J Cardiol 36:453–458
129. Cintron G (1987) Evaluation of intravenous isosorbide dinitrate and nitroglycerin in the treatment of acute myocardial infarction with elevated pulmonary capillary wedge pressure. Eur Heart J 8 (Suppl 1):41
130. Judgutt BI, Becker LC, Hutchins GM, Bulkley BH, Reid PR, Kallman CH (1981) Effect of intravenous nitroglycerin on collateral blood flow and infarct size in the conscious dog. Circulation 63:17–28
131. Flaherty JT, Reid PR, Kelly DT, Taylor DR, Weisfeldt ML, Pitt B (1975) Intravenous nitroglycerin in acute myocardial infarction. Circulation 51:132–139
132. Borer JS, Redwood DR, Levitt B et al (1975) Reduction in myocardial ischemia with nitroglycerin or nitroglycerin plus phenylephrine adminstered during acute myocardial infarction. N Engl J Med 293:1008–1012

133. Epstein SE, Kent KM, Goldstein RE et al (1975) Reduction of ischemic injury by nitroglycerin during acute myocardial infarction. N Engl J Med 292:29–35
134. Smith ER, Redwood DR, McCarron WE, Epstein SE (1973) Coronary artery occlusion in the conscious dog. Effects of alterations in arterial pressure produced by nitroglycerin, hemorrhage and alpha-adrenergic agonists on the degree of myocardial ischemia. Circulation 47:51–57
135. Myers RW, Scherer JL, Goldstein RA, Goldstein RE, Kent KM, Epstein SE (1975) Effects of nitroglycerin and nitroglycerin-methoxamine during acute myocardial ischemia in dogs with pre-existing multivessel coronary occlusive disease. Circulation 51:632–640
136. Hirshfeld JW Jr, Borer JS, Goldstein RE, Barrett MJ, Epstein SE (1974) Reduction in severity and extent of myocardial infarction when nitroglycerin and methoxamine are adminstered during coronary occlusion. Circulation 49:291–297
137. Come PC, Flaherty JT, Baird MG et al (1975) Reversal by phenylephrine of the beneficial effects of intravenous nitroglycerin in patients with acute myocardial infarction. N Engl J Med 293:1003–1007
138. Chiche P, Derrida JP (1976) Cartographie précordiale dans l'infarctus myocardique récent. Nouv Presse 5:3007–3009
139. Chiche P, Derrida JP, Baligadoo S (1979) Effets cliniques et hémodynamiques de la nitroglycerine intraveineuse dans le traitement de l'infarctus du myocarde de l'insuffisance cardiaque. Thérapie 34:685–704
140. Derrida JP, Sal R, Chiche P (1978) Effects of prolonged nitroglycerin infusion in patients with acute myocardial infarction. Am J Cardiol 41:407
141. Bussmann WD, Passek D, Seidel W, Kaltenbach M (1981) Reduction of CK and CK-MB indexes of infarct size by intravenous nitroglycerin. Circulation 63:615–622
142. Kim YI, Williams JF Jr (1982) Large dose sublingual nitroglycerin in acute myocardial infarction: relief of chest pain and reduction of Q wave evolution. Am J Cardiol 49:842–848
143. Jaffe AS, Geltman EM, Tiefenbrunn AJ et al (1983) Reduction of infarct size in patients with inferior infarction with intravenous glyceryl trinitrate. Br Heart J 49:452–460
144. Flaherty JT, Becker LC, Bulkley BH et al (1983) A randomized prospective trial of intravenous nitroglycerin in patients with acute myocardial infarction. Circulation 68:576–588
145. Sheehan FH, Goldstein RE, Bolli R, Epstein SE (1983) The effects of nitroglycerin-methoxamine combination on infarct size in conscious dogs. Am Heart J 105:37–43
146. Grosser KD, Schuster P, Freund M, Knoch K, Sturm A (1983) Oral ISDN for the treatment of patients with acute myocardial infarction. Z Kardiol 72 (Suppl 3):147–151
147. Sclarovsky S, Klainman E, Lewin RF, Arditti A, Agmon J. The effect of high doses of isosorbide dinitrate within a short time on the very early electrocardiographic stages of acute anterior wall myocardial infarction. Am J Cardiol (Suppl) 61 (9):78E–80E
147a. Jugdutt BI, Warnica JW (1988) Intravenous nitroglycerin therapy to limit myocardial infarct size, expansion, and complications. Circulation 78:906–919
148. Hackett D, Davies G, Chierchia S, Maseri A (1987) High frequency of intermittent coronary occlusion in acute myocardial infarction – role of ISDN. Eur Heart J 8 (Suppl 1):35
149. Flaherty JT (1987) The role of nitrates in acute myocardial infarction. Eur Heart J 8 (Suppl 1):38
150. Bussmann WD (1981) The role of nitroglycerin in acute myocardial infarction. In: Nitrates III. Lichtlen PR, Engel H-J, Schrey A, Swan HJC (eds) Springer-Verlag, Berlin 329–340
150a. Lam JYT, Chesebro JH, Fuster V (1988) Platelets, vasoconstriction, and nitroglycerin during arterial wall injury. Circulation 78:712–716

151. Guiha NH, Cohn JN, Mikulic E, Franciosa JA, Limas CJ (1974) Treatment of refractory heart failure with infusion of nitroprusside. N Engl J Med 291:587–592
152. Franciosa JA, Blank R, Cohn JN, Mikulic E (1977) Hemodynamic effects of topical, oral and sublingual nitroglycerin in left ventricular failure. Curr Ther Res 22:231–245
153. Cohn JN, Mathew KJ, Franciosa JA, Snow JA (1974) Chronic vasodilator therapy in the management of cardiogenic shock and intractable left ventricular failure. Ann Intern Med 81:777–780
154. Franciosa JA, Mikulic E, Cohn JN, Jose E, Fabie A (1974) Hemodynamic effects of orally administered isosorbide dinitrate in patients with congestive heart failure. Circulation 50:1020–1024
155. Mikulic E, Franciosa JA, Cohn JN (1975) Comparative hemodynamic effects of chewable isosorbide dinitrate and nitroglycerin in patients with congestive heart failure. Circulation 52:477–482
156. Rabinowitz B, Tamari I, Elazar E, Neufeld NH (1982) Intravenous isosorbide dinitrate in patients with refractory pump failure and acute myocardial infarction. Circulation 65:771–778
157. Baligadoo S, Ingrand JC, Derrida JP, Savier CH, Chiche P (1978) Etude des effets hémodynamiques de la trinitrine intraveineuse. Coeur Med Int 17:231–242
158. Williams DO, Bommer WJ, Miller RR, Amsterdam EA, Mason DT (1977) Hemodynamic assessment of oral peripheral vasodilator therapy in chronic congestive heart failure: prolonged effectiveness of isosorbide dinitrate. Am J Cardiol 39:84–90
159. Stephens JD, Camm AJ, Spurrell RAJ (1978) Improvement in exercise hemodynamics by isosorbide dinitrate in patients with severe congestive cardiac failure secondary to ischemic heart disease. Br Heart J 40:832–837
160. Bussmann W, Lohner J, Kaltenbach M (1977) Orally administered isosorbide dinitrate in patients with and without left ventricular failure due to acute myocardial infarction. Am J Cardiol 39:91–96
161. Rubin SA, Gelberg HJ, Swan HJC (1981) Attenuation of nitrate preload reduction by exercise in patients with severe chronic heart failure. In: Nitrates III. Lichtlen PR, Engel H-J, Schrey A, Swan HJC (eds) Springer-Verlag, Berlin 406–410
162. Leier CV, Magorien RD, Desch CE, Thompson MJ, Unverferth DV (1981) Hydralazine and isosorbide dinitrate: Comparative central and regional hemodynamic effects when administered alone or in combination. Circulation 63:102–109
163. Franciosa JA, Cohn JN (1979) Effect of isosorbide dinitrate on response to submaximal and maximal exercise in patients with congestive heart failure. Am J Cardiol 43:1009–1014
164. Franciosa JA, Goldsmith SR, Cohn JN (1980) Contrasting immediate and long-term effects of isosorbide dinitrate on exercise capacity in congestive heart failure. Am J Med 69:559–566
165. Wilson JR, Ferraro N (1981) Circulatory improvement after hydralazine or isosorbide dinitrate administration in patients with heart failure. Am J Med 71:627–633
166. Hayes PC, Morrison L, Bouchier IAD (1985) Effect of glyceryl trinitrate, isosorbide-5-mononitrate and propranolol on apparent liver blood flow In: Mononitrates. Cohn JN, Rittinghausen R (eds) Springer-Verlag, Berlin 73–77
167. Hecht HS, Karahalios SE, Schnugg SJ et al (1982) Improvement in supine bicycle exercise performance in refractory congestive heart failure after isosorbide dinitrate: radionuclide and hemodynamic evaluation of acute effects. Am J Cardiol 49:133–140
168. Marmor A, Blich M, Schneeweiss A (1988) Very early hemodynamic improvement with isosorbide dinitrate spray in patients with acute pulmonary edema. Am J Cardiol 9:4E–6E
169. Bussmann WD, Schupp D (1978) Effect of sublingual nitroglycerin in emergency treatment of severe pulmonary edema. Am J Cardiol 41:931–936
170. Magrini F, Niarchos AP (1980) Ineffectiveness of sublingual nitroglycerin in acute left ventricular failure in the presence of massive peripheral edema. Am J Cardiol 45:841–847

171. Armstrong PW, Armstrong JA, Marks GS (1980) Pharmacokinetic-hemodynamic studies of intravenous nitroglycerin in congestive cardiac failure. Circulation 62:160–166
172. Armstrong PW, Walker DC, Burton JP, Parker JO (1975) Vasodilator therapy in acute myocardial infarction. A comparison of sodium nitroprusside and nitroglycerin. Circulation 52:1118–1122
173. Strauer BE, Scherpe A (1978) Ventricular function and coronary hemodynamics after intravenous nitroglycerin in coronary artery disease. Am Heart J 95:210–219
174. Bonoron-Adèle S, Tariosse L, Bricaud H, Besse P (1981) Demonstration of a real inotropic effect of nitrites on myocardial contractility during anoxia and reoxygenation: Attempt to determine action site with the contribution of calcium antagonist compounds. In: Nitrates III. Lichtlen PR, Engel H-J, Schrey A, Swan HJC (eds) Springer-Verlag, Berlin, 157–169
175. Khatri L, Uemura N, Notargiacomo A, Freis ED (1977) Direct and reflex cardiostimulating effects of hydralazine. Am J Cardiol 40:38–42
176. Leier CV, Desch CE, Magorien RD et al (1980) Positive inotropic effects of hydralazine in human subjects: comparison with prazosin in the setting of congestive heart failure. Am J Cardiol 46:1039–1044
176a. Obata K, Yasue H, Okumura K, Kurose M, Ogawa H, Imura H, Nakao K (1989) Effects of isosorbide dinitrate (ISDN) on plasma atrial natriuretic polypeptide (ANP) level in patients with congestive heart failure (CHF). Cardiovasc Drugs & Ther 3 (Suppl 2):617 abstr
177. Borer JS, Redwood DR, Itsoita SB, Goldstein RE, Epstein SE (1978) Nitroglycerin-induced improvement in exercise tolerance and hemodynamics in patients with chronic rheumatic heart valve disease. Am J Cardiol 41:302–307
178. Klein G, Wirtzfeld A, Delius W, Schnelle K, Schinz A (1980) Nitroglycerin treatment of pulmonary congestion in decompensated valvular disorders. In: Nitrates II. Rudolph W, Schrey A (eds) Urban & Schwarzenberg, Munich 354–368
179. Goppel L, Dacian S, Henze E, Rudolph W (1980) Effect of sublingual isosorbide dinitrate on left ventricular contractility, volume, and wall stress in aortic regurgitation. In: Nitrates II Rudolph W, Schrey A (eds) Urban & Schwarzenberg, Munich 369–375
180. Baxley WA, Schrank JP, Reynolds DW (1980) Acute hemodynamic effects of sublingual nitroglycerin in patients with valvular insufficiency. In: Nitrates II. Rudolph W, Schrey A (eds) Urban & Schwarzenberg, Munich 385–396
181. Roth A, Kumar A, Kulick D, McKay CR, Kawanishi D, Rahimtoola SH (1986) Beneficial effect of nitrate therapy on left ventricular hemodynamics and function in patients with chronic severe mitral regurgitation. J Am Coll Cardiol 7 (Suppl):170A
182. Kawashita K, Kambara H, Kadota K, Saimyoji H, Kawai C (1983) Radiocardiographic assessment of dobutamine and isosorbide dinitrate therapy in patients with mitral stenosis and pulmonary congestion. Jpn Circ J 47:283–288
183. Tebbe U, Sauer G, Neuhaus KL, Kreuzer H (1981) Effects of nitroglycerin on left ventricular and myocardial function in aortic stenosis. In: Nitrates III. Lichtlen PR, Engel H-J, Schrey A, Swan HJC (eds) Springer-Verlag, Berlin 558–564
184. Tebbe U, Neuhaus L, Sauer G, Neumann P, Kreuzer H (1983) Nitrates in aortic valve disease: Acute and chronic effects. Z Kardiol 72 (Suppl 3):152–155
184a. Klepzig HH, Warner KG, Siouffi SY, Saad AJ, Hayes A, Kaltenbach M, Khuri SF (1989) Hemodynamic effects of nitroglycerin in an experimental model of acute aortic regurgitation. J Am Coll Cardiol 13:927–935
184b. Chatterjee K, Parmley WW (1977) The role of vasodilator therapy in heart failure. Prog Cardiovasc Dis 19:301–325
184c. Miller RR, Vismara LA, DeMaria AN, Salel AF, Mason DT (1976) Afterload reduction therapy with nitroprusside in severe aortic regurgitation: improved cardiac performance and reduced regurgitant volume. Am J Cardiol 38:564–567
185. Chandraratna PAN, Langevin E, O'Dell R, Rubenstein C, San Pedro S (1978) Use of nitroglycerin ointment in congestive heart failure. Cardiology 63:337–342

186. Gray R, Chatterjee K, Vyden JK, Ganz W, Forrester JS, Swan HJC (1975) Hemodynamic and metabolic effects of isosorbide dinitrate in chronic congestive heart failure. Am Heart J 90:346–352
186a. Natarajan D, Khurana TR, Karhade V, Nigam PD (1988) Sustained hemodynamic effects with therapeutic doses of intravenous nitroglycerin in congestive heart failure. Am J Cardiol 62:319–321
186b. Packer M, Lee WH, Kessler PD, Gottlieb SS, Medina N, Yushak M (1987) Prevention and reversal of nitrate tolerance in patients with congestive heart failure. N Engl J Med 317:799–809
187. Klütsch K, Schöniger F, Koch E, Grosswendt J (1985) Der Stellenwert der Nitrate in der Therapie der chronischen Herzinsuffizienz. In: Mononitrat. Borchard U, Rafflenbeul W, Schrey A (eds) Universitätsdruckerei und Verlag Dr C Wolf und Sohn, München 126–133
188. Schelling J, Lasagna L (1967) A study of cross tolerance to circulatory effects of organic nitrates. Clin Pharmacol ther:256–260
189. Needleman P (1971) Tolerance to the vascular effects of glyceryl trinitrate. J Pharmacol Exp Ther 171:98–102
190. Rush MI, Lang WJ, Rand MJ (1971) Studies on compensatory reflexes and tolerance to glyceryl trinitrate (GTN). Eur J Pharmacol 16:148–155
191. Lee G, Mason DT, DeMaria AN (1978) Effects on long-term oral administration of isosorbide dinitrate on the antianginal response to nitroglycerin. Am J Cardiol 41:82–87
192. Distante A, L'Abbate A, Palombo C (1981) May prolonged high doses of nitrates cause tolerance? Preliminary results on the response to an additional dose by infusion. In: Nitrates III. Lichtlen PR, Engel H-J, Schrey A, Swan HJC (eds) Springer-Verlag, Berlin 82–90
193. Abrams J (1980) Nitrate tolerance and dependence. Am Heart J 99:113–123
194. Hall D, Stautner C, Rudolph W (1986) Intermittent treatment of congestive heart failure with nitrates: pharmacodynamic basis for circumvention of tolerance development. Circulation 74 (Suppl II) II-508 (2028)
194a. Makhoul N, Dakak N, Flugelman MY, Merdler A, Shefer A, Schneeweiss A, Halon DA, Lewis BS (1989) Nitrate tolerance in heart failure: differential venous, pulmonary and systemic effects? Eur Heart J 10 (Supl):72 abstr
194b. Packer M, Kessler PD, Gottlieb SS, Lee WH, Kukin ML, Medina M, Yushak M (1988) Does neurohormonal activation contribute to the development of nitrate tolerance in patients with chronic heart failure? Circulation 78 (Suppl II):II-27 (0106)
195. Francis GS, Archibald DG, Cohn JN (1986) Mortality reduction by hydralazine – isosorbide dinitrate therapy in congestive heart failure: -the V-HeFT study. Circulation 74 (Suppl II):II-508 (2025)
195a. Cohn JN, Johnson C, Cooperative Study Group (1988) Mechanism of death and its relationship to baseline variables and treatment in V-HeFT. Circulation 78 (Suppl II):II-347 (1383)
195b. Fletcher R, Johnson G, Cohn J and VHEFT VA Cooperative Study Group (1988) Vasodilator improvement of mortality in CHF is dependent on the presence of ventricular arrhythmia. Circulation 78 (Suppl II):II-347 (1384)
196. Walsh WF, Greenberg BH (1981) Results of long-term vasodilator therapy in patients with refractory congestive heart failure. Circulation 64:499–505
197. Turner GG, Weir EK, Chesler R, Pierpont GL (1981) Reassessment of vasodilator therapy in angina: effects of oral isosorbide dinitrate and hydralazine on exercise tolerance in patients receiving propranolol therapy. Am J Cardiol 47:910–916
198. Massie B, Chatterjee K, Werner J, Greenberg B, Hart R, Parmley WW (1977) Hemodynamic advantage of combined administration of hydralazine orally and nitrates nonparenterally in the vasodilator therapy of chronic heart failure. Am J Cardiol 40:794–801
199. Franciosa JA, Cohn JN (1979) Immediate effects of hydralazine-isosorbide dinitrate combination on exercise capacity and exercise hemodynamics in patients with left ventricular failure. Circulation 59:1085–1091

200. Franciosa JA, Weber KT, Levine TB et al (1982) Hydralazine in the long-term treatment of chronic heart failure: Lack of difference from placebo. Am Heart J 104:587–594
201. Lewis BS, Halon DA, Hardoff R. Effects of captopril and isosorbide dinitrate on LV function and myocardial contractility in severe cardiac failure. Am J Cardiol (in press)
201a. Haude M, Steffen W, Pop T, Tschollar W, Meyer J (1988) Sublingual captopril versus nitroglycerine in patients with left heart failure – a randomized cross-over study. Circulation 78 (Suppl II):II-618 (2464)
202. Bussmann WD (1987) Treatment of cardiogenic shock with low dose nitroglycerin. Eur Heart J 8 (Suppl 1):45
203. Bertel O, Noll G (1981) Effects of N-acetylcysteine on nitroglycerin responsiveness before and during nitrate therapy of congestive heart failure. Eur Heart J 8 (Suppl 1):44
204. Massie BM, Kramer B, Shen E, Haughom F (1981) Vasodilator treatment with isosorbide dinitrate and hydralazine in chronic heart failure. Br Heart J 45:376–384
205. Packer M, Medina N, Yushak M (1983) Identification of four different hemodynamic patterns of response to captopril therapy in severe heart failure. J Am Coll Cardiol 1:727
206. Packer M, Medina V, Yushak RN, Lee W-H (1985) Comparative effects of captopril and isosorbide dinitrate on pulmonary arteriolar resistance and right ventricular function in patients with severe left ventricular failure: Results of a randomized crossover study. Am Heart J 109:1293–1299
207. Nelson GIC, Silke B, Ahuja RC, Hussain M, Taylor SH (1983) Management of acute heart failure following myocardial infarction. Hemodynamic advantages of isosorbide dinitrate over frusemide. Z Kardiol 72 (Suppl 3):141–146
208. Hutton I, Martin W, McGhie I, Tweddel AC (1987) A comparison of intravenous isosorbide-5-mononitrate and frusemide on coronary and systemic haemodynamics in patients with chronic cardiac failure. Eur Heart J 8 (Suppl):43
209. Caesar K, Wegener B, Hossmann K (1980) Effect of isosorbide dinitrate on blood pressure, peripheral vascular resistance and venous capacity in patients with arterial hypertension. In: Nitrates II. Rudolph W, Schrey A (eds) Urban & Schwarzenberg, Munich, 135–142
210. Duchier J, Iannascoli M, Safar ME (1986) Antihypertensive effect of sustained-release isosorbide dinitrate in elderly patients with systolic hypertension. X World Congress of Cardiology, Washington, DC, September 14–19 (2796)
210a. Fitchett DH, Simkus GJ, Beaudry JP, Marpole DG (1988) Reflected pressure waves in the ascending aorta: effect of nitroglycerin. Circulation 78 (Suppl II):II-105 (0417)
211. Häge R (1985) Untersuchung über die antiarrhythmische Wirkung von IS-5-N – Doppelblindstudie cross-over zu Verapamil. In: Mononitrat. Hochrein T, Tauchert M, Schrey A, Weiss M (eds) Universitätsdruckerei und Verlag Dr C Wolf und Sohn, München, 75–86
212. Senges J, Zebe H, Pelzer D, Brachmann J, Krämer B, Kübler W (1980) "Anitarrhythmic" effect of nitrates: Clinical and experimental studies. In: Nitrates II. Rudolph W, Schrey A (eds) Urban & Schwarzenberg, Munich, 87–99
213. Discussion Paper Senges (1980) In: Nitrates II. Rudolph W, Schrey A (eds) Urban & Schwarzenberg, Munich, 100
214. Roberts AJ, Niarchos AP, Subramanian VA et al (1977) Systemic hypertension associated with coronary artery bypass surgery. J Therap Cardiovasc Surg 74:846–859
215. Whelton PK, Flaherty T, MacAllister NP et al (1980) Hypertension following coronary artery bypass surgery: The role of preoperative propranolol therapy. Hypertension 2:291–298
216. Kaplan JA, Welss PH, Hug CC, Holland CE (1980) Nitroglycerin: Effects of ventricular function and bypass. Anesthesiology 53 (Suppl):83
217. Zadrobilek E, Draxler V, Hammerle A, Schindler I (1983) Hemodynamic and hormonal response to induced hypotension with isosorbide dinitrate and nitroglycerin during anesthesia. Z Kardiol 72 (Supl 3):156–161

218. Kober G, Grossmann R, Schulz W, Kaltenbach M (1981) Effect of nitroglycerin on the diameter of peripheral arterial and venous pulmonary vessels. In: Nitrates III. Lichtlen PR, Engel H-J, Schrey A, Swan HJC (eds) Springer-Verlag, Berlin 280–284
219. Bonoron-Adèle S, Tariosse L, Bricaud A, Besse P (1981) Demonstration of a real inotropic effect of nitrites on myocardial contractility during anoxia and reoxygenation: attempt to determine action site with the contribution of calcium antagonist compounds. In: Nitrates III. Lichtlen PR, Engel- H-J, Schrey A, Swan HJC (eds) Springer-Verlag, Berlin 157–169
220. Brent BN, Berger HJ, Matthay RA, Mahler D, Pytlik L, Zaret BL (1983) Contrasting acute effects of vasodilators (nitroglycerin, nitroprusside, and hydralazine) on right ventricular performance in patients with chronic obstructive pulmonary disease and pulmonary hypertension: a combined radionuclide-hemodynamic study. Am J Cardiol 51:1682–1689
221. Mathys H, Rühle KH, Haller T (1981) Nitrates in patients with pulmonary hypertension due to airways obstruction. In: Nitrates III. Lichtlen PR, Engel H-J, Schrey A, Swan HJC (eds) Springer-Verlag, Berlin 285–292
222. Ježek V, Zežková J, Michaljanič A, Fučik J (1987) Long-term reduction of pulmonary hypertension in the interstitial lung fibrosis by isosorbide dinitrate. Eur Heart J 8 (Suppl 1):51
223. Hirshleifer I, Arora Y (1961) Nitrates in the treatment of bronchial asthma. Chest 39:275–283
224. Okayama M, Sasaki H, Takishima T (1984) Bronchodilator effect of sublingual isosorbide dinitrate in asthma. Eur J Clin Pharmacol 26:151–155
225. Niehues B, Jansen W, Oberhoffer H, Thoma R, Kupper JH, Behrenbeck DW (1981) Comparison of the effects of nitroglycerin and aminophylline on hemodynamics and lung function in patients with chronic obstructive lung disease. In: Nitrates III. Lichtlen PR, Engel H-J, Schrey A, Swan HJC (eds) Springer-Verlag, Berlin 293–298
226. Miller WC, Shultz TF (1979) Failure of nitroglycerin as a bronchodilator (letter). Am Rev Respir Dis 120:471
227. Kennedy T, Summer WR, Sylvester J, Robertson D (1981) Airway response to sublingual nitroglycerin in acute asthma. JAMA 246:145–147
228. Draxler V, Mauritz W, Sporn P (1981) Influence of isosorbide dinitrate on pulmonary circulation and gas exchange in acute respiratory distress syndrome. In: Nitrates III. Lichtlen PR, Engel H-J, Schrey A, Swan HJC (eds) Springer-Verlag, Berlin 314–322
229. Schneider W, Tessmer G, Strohm WD, Kober K, Kaltenbach M (1984) Diameter of abdominal vessels during acute and sustained nitrate therapy. Eur Heart J 5 (Suppl 1):28
230. Lund f (1948) Percutaneous nitroglycerin treatment in cases of peripheral circulatory disorders, especially Raynaud's disease. Acta Med Scand 206 (Suppl):196–206
231. Francis DR, Hubbard ER, Johnson LE (1977) Nitroglycerin ointment as a vasodilator in the lower extremities. J Am Podiatry Assoc 67:874–879
231a. Testa R, Biagini A, Michelassi C, Emdin M, Mazzei MG, Carpeggiani C, L'Abbate A. Improvement of walking distance in patients with intermittent claudication by chronic local therapy with isosorbide dinitrate ointment. Angiology (in press)
232. Biagini A, Carpeggiani C, Testa R, Mazzei MG, Michelassi C, L'Abbate A (1981) Claudicatio intermittens: trattamento topico con isosorbide dinitrato in pasta. Risultati preliminari. G Ital Cardiol 11:983–989
233. Gelfond M, Rozen P, Keren S, Gilat T (1981) Effect of nitrates on LES pressure in achalasia: a potential therapeutic aid. Gut 22:312–318
234. Rozen P, Gelfond M, Salzman S, Baron J, Gilat T (1982) Radionuclide confirmation of the therapeutic value of ISDN in relieving the dysphagia in achalasia. J Clin Gastroenterol 4:17–22
235. Gelfond M, Rozen P, Gilat T (1982) Isosorbide dinitrate and nifedipine treatment of achalasia: a clinical, manometric and radionuclide evaluation. Gastroenterol 83:963–969

236. Freeman JG, Barton JR, Record CO (1985) Effect of isosorbide dinitrate, verapamil, and labetalol on portal pressure in cirrhosis. Br Med J 291:561–562
237. Hallemans R, Naeije R, Mols P, Mélot C, Reding P (1983) Treatment of portal hypertension with ISDN alone and in combination with vasopressin. Crit Care Med 11:536–540
238. Withworth CG, Grant WM (1964) Use of nitrate and nitrite vasodilators by glaucomatous patients. Arch Ohthalmol 71:492–496
239. Wizemann AJS, Wizemann V (1980) Organic nitrate therapy in glaucoma. Am J Ophthalmol 90:106–109
240. Wizemann A (1982) Zweijährige Erfahrungen mit peroraler Nitrattherapie des Glaukoms. Fortschr Ophthalmol 79:149–150
241. Ilbawi MN, Idriss FS, De Leon SY, Berry TE, Paul MH (1984) Hemodynamic effects of intravenous nitroglycerin in postoperative pediatric cardiac patients. Circulation 70 (Suppl II, 99):II 25
241a. Marchionni N, Schneeweis A, Di Bari M et al (1988) Age-related hemodynamic effects of intravenous nitroglycerin for acute myocardial infarction and left ventricular failure. Am J Cardiol (Suppl) 61 (9):81E–83E
241b. Gascho JA, Fanelli C, Zelis R (1989) Aging reduces venous distensibility and the venodilatory response to nitroglycerin in normal subjects. Am J Cardiol 63:1267–1270
242. Fleisch JH, Hooker CS (1976) The relationship between age and relaxation of vascular smooth muscle in the rabbit and rat. Circ Res 38:243–249
243. Figueras J, Taylor WR, Ogawa T, Forrester JS, Singh BM, Swan HJC (1979) Comparative haemodynamic and peripheral vasodilator effects of oral and chewable isosorbide dinitrate in patients with refractory congestive cardiac failure. Br Heart J 41:317–324
244. Lin SG, Flaherty JT (1985) Crossover from intravenous to transdermal nitroglycerin therapy in unstable angina pectoris. Am J Cardiol 56:742–748
245. Willis WH, Russell RO, Mantle JA (1974) The hemodynamic response to sublingual isosorbide dinitrate in unstable angina pectoris. Am J Cardiol 33:179 abstr.
246. Rudolph W, Dacian S, Dirschinger J, Fleck E,Loracher C, Redl A (1976) Der Einfluß von Nitraten auf regionale Myocarddurchblutung und regionale Ventrikelwandbewegung. Med Klin 71:1539–1545
247. Brown BG (1983) Dilatation of coronary artery stenosis: A major mechanism of the effect of nitrates. Z Kardiol 72 (Suppl 3):77–81
248. Sweatman T, Strauss G, Selzer A, Cohn KE (1972) The long acting hemodynamic effects of isosorbide dinitrate. Am J Cardiol 29:475–480
249. Baxter RH, Tait CM, McGuinness JB (1977) Vasodilator therapy in acute myocardial infarction. Use of sublingual isosorbide dinitrate. Br Heart J 39:1067–1070
250. Kiely J, Kelly DT, Taylor DR, Pitt B (1973) The role of frusemide in the treatment of left ventricular dysfunction associated with acute myocardial infarction. Circulation 48:581–587
251. Armstrong PW, Armstrong JA, Marks GS (1979) Blood level after sublingual nitroglycerin. Circulation 59:585–588
252. Morrison RA, Wiegand UW, Jähnchen DE et al (1983) Isosorbide dinitrate kinetics and dynamics after intravenous, sublingual, and percutaneous dosing in angina. Clin Pharmacol Ther 33:747–756
253. Doyle E, Chasseaud LF, Taylor T (1980) Measurement of plasma concentrations of isosorbide dinitrate. Biopharm Drug Dispos 1:141–147
254. Robinson BF (1968) Mode of action of nitroglycerin in angina pectoris. Correlation between haemodynamic effects during exercise and prevention of pain. Br Heart J 30:295–302
255. Sandler G, Ilani MA, Lawson CW (1963) Glyceryl trinitrate in angina pectoris. Lancet 1:1130–1136
256. Kattus AA, Alvaro AB, Coulson A (1975) Effectiveness of isosorbide dinitrate and nitroglycerin in relieving angina pectoris during uninterrupted exercise. Chest 67:640–646

257. Detry JR, Bruce RA (1971) Effect of nitroglycerin on "maximal" oxygen intake and exercise electrocardiogram in coronary heart disease. Circulation 43:155–163
257a. Toman J, Lupinek Z, Janousek S, Rotrekl P, Seda L, Soucek M (1990) Comparison of hemodynamic effects of nitroglycerin tablets and isosorbide dinitrate spray in patients with acute myocardial infarction. C.V. World Report (in press)
258. Kimchi A, Lee G, Amsterdam E, Fujii K, Krieg P, Mason DT (1983) Increased exercise tolerance after nitroglycerin oral spray: a new and effective therapeutic modality in angina pectoris. Circulation 67:124–127
259. Rouleau J, Boerboom LE, Surjadhana A, Hoffmann JIE (1979) The role of autoregulation and tissue diastolic pressures in the transmural distribution of left ventricular blood flow in anesthetized dogs. Circ Res 45:804–815
260. Capurro NL, Kent KM, Epstein SE (1977) Comparison of nitroglycerin-, nitroprusside-, and phentolamine-induced changes in coronary collateral function in dogs. J Clin Invest 60:295–301
261. Taylor T, Chasseaud LF, Doyle E (1980) Pharmacokinetics of isosorbide dinitrate after intravenous infusion in human subjects. Bioharm Drug Dispos 1:149–156
262. Fung H-L, Parker JO (1983) Prolonged plasma half-life after oral isosorbide dinitrate in patients with angina pectoris. Br J Clin Pharmacol 15:746–748
263. Schneeweiss A, Blich M, Marmor A. Reduced duration of relief of ischemia with isosorbide dinitrate spray: an improved therapeutic modality in angina pectoris (unpublished data)
264. Wildfeurer A, Fries W, Laufen H (1982) Aspekte zur pharmazeutischen Entwicklung eines neuen oralen Isosorbidnitrat-Sprays. Therapiewoche 32:6141–6143
265. Grüntzig A (1980) Onset of effect of isosorbide dinitrate spray. In: Nitrates II. Rudolph W, Schrey A (eds) Urban & Schwarzenberg, Munich 847–854
266. Feldman RL, Marx JD, Pepine CJ, Conti CR (1982) Analysis of coronary responses to various doses of intracoronary nitroglycerin. Circulation 66:321–327
267. Kern MJ, Miller JT, Henry RL (1986) Dose-related effects of intracoronary nitroglycerin on coronary hyperemia in patients with coronary artery disease. Am Heart J 111:845–852
268. Hood WP, Amende I, Simon R, Lichtlen PR (1980) The effects of intracoronary nitroglycerin on left ventricular systolic and diastolic function in man. Circulation 61:1098–1104
269. Rutsch W, Schmutzler H (1981) Left ventricular function and regional wall motion after intracoronary application of nitrates in coronary heart disease. In: Nitrates III. Lichtlen PR, Engel H-J, Schrey A, Swan HJC (eds) Springer-Verlag, Berlin 134–140
270. Ganz W, Marcus HS (1976) Failure of intracoronary nitroglycerin to alleviate pacing-induced angina. Circulation 46:880–889
271. Winniford MD, Jackson J, Mallory CR, Rehr RB, Campbell WB, Hillis LD (1984) Does indomethacin attenuate the coronary vasodilatory effect of nitroglycerin? J Am Coll Cardiol 4:1114–1117
272. Liu P, Houle S, Burns RJ (1985) Effect of intracoronary nitroglycerin on myocardial blood flow and distribution in pacing-induced angina pectoris. Am J Cardiol 55:1270–1276
273. Miller JT, Henry RL, Kern MJ, O'Rourke RA (1968) The effect of intracoronary nitroglycerin on myocardial blood flow to coronary collateral-dependent regions. J Am Coll Cardiol 7:230A abstr
274. Erbel R, Schreiner G, Henkel B, Pop T, Meyer J (1983) Improved ischemic tolerance during percutaneous transluminal coronary angioplasty by intracoronary injection of nitroglycerin. Z Kardiol 72 (Suppl 3):71–73
275. Hermann G, Simon R, Lichtlen PR (1985) Intracoronary versus intravenous nitroglycerin during coronary angioplasty. Eur Heart J 6 (Suppl 1):24
276. Crake T, Quyyum AA, Rickards AF, Fox KM (1987) The effects of intraocoronary isosorbide dinitrate during coronary angioplasty. Eur Heart J 8 (Suppl 1):25
277. Kern M, Vandormael M, Deligonul U, Gibson P, Harper M, Presant S (1986) Attenuation of intracoronary nitroglycerin-induced coronary vasodilatory responses after percutaneous angioplasty in patients. Circulation 74 (Suppl II):539

278. Udhoji VN, Heng MK (1984) Hemodynamic effects of high-dose sustained-action oral isosorbide dinitrate in stable angina. Am J Med 76:234–240
279. Flugelman MY, Shefe RA, David YB, Witt H, Gotsman MS (1985) Effect of sustained-release isosorbide dinitrate on exercise performance. Cardiology 72:123–128
280. Ohlmeier H, Mertens HM, Moller M, Mannebach H, Gleichmann U (1986) Vereinfachte Langzeittherapie der koronaren Herzkrankheit mit 120 mg retardiertem Isosorbidnitrat einmal täglich. Untersuchung zur Wirkdauer und Toleranzentwicklung. Z Kardiol 75 (Suppl 3):50
281. Boertz A, Bonn R (1986) Nitrate therapy without loss of action by correct dosage. Z Kardiol 75 (Suppl 3):57–60
282. Schneider W, Kaltenbach M (1986) Dosiswirkungsbeziehungen bei der Therapie der Angina pectoris mit Nitraten. Z Kardiol 75 (Suppl 3):61–67
283. Wortmann A, Bachmann K, Deuber HJ, Roth E (1986) Haemodynamik und Koronardynamik unter ISDN-Langzeittherapie. Z Kardiol 75 (Suppl 3):68–76
284. Dietmann K, Sponer G, Voss E (1981) Pharmakodynamik, Pharmakokinetik und Metabolismus der Nitrate des Isosorbids am wachen Hund. Med Welt 32:481–491
285. Sponer G, Kühnle HF, Strein K, Bartsch W, Endele R, Dietmann K (1983) Pharmacokinetic aspects of isosorbide-5-mononitrate in dogs. J Pharmacol Exp Ther 228:235–239
286. Abshagen U, Betzien G, Endele R, Kaufmann B (1981) Pharmacokinetics of intravenous and oral isosorbide-5-mononitrate. Eur J Clin Pharmacol 20:269–275
287. Abshagen U, Spörl-Radun S (1981) First data on effects and pharmacokinetics of isosorbide-5-mononitrate in normal man. Eur J Clin Pharmacol 19:423–429
288. Strein K, Bartsch W, Sponer G, Müller-Beckmann B, Lexa P (1984) Differences in the nitrate ion formation and the toxicological findings between isosorbide dinitrate and isosorbide-5-mononitrate. Toxikol Appl Pharmacol 72:142–147
289. Seidel F, Michel D (1981) Comparative haemodynamic and pharmacokinetic investigations after oral isosorbide-2-mononitrate and isosorbide-5-mononitrate. In: Nitrates III. Lichtlen PR, Engel H-J, Schrey A, Swan HJC (eds) Springer-Verlag, Berlin 54–60
290. Krepp HP (1985) Langzeitbehandlung der koronaren Herzerkrankung mit elantan long. In: Mononitrat. Borchard U, Rafflenbeul W, Shrey A (eds) Universitätsdruckerei und Verlag Dr C Wolf und Sohn, München, 176–183
291. Eggeling T, Osterspey A, Geppert R, Jansen W (1985) Antianginöse Wirksamkeit von retardiertem Isosorbid-5-Mononitrat bei Patienten mit koronarer Herzkrankheit. In: Mononitrat. Borchard U, Rafflenbeul W, Schrey A (eds) Universitätsdruckerei und Verlag Dr C Wolf und Sohn, München, 76–87
292. Akhras F, Jackson G (1986) Isosorbide-5-mononitrate as monotherapy in patients with chronic stable angina. J Am Coll Cardiol 7:230A abstr
293. Bidoggia JH, Casar JE, Hirschon AJ, Campo A, Nordaby R (1985) Beneficial effects of isosorbide-5-mononitrate on exercise performance in patients with chronic stable angina pectoris. In: Mononitrates. Cohn JN, Rittinghausen R (eds) Springer-Verlag, Berlin, 201–214
294. Schiavoni G, Montenero AS, Mazzari M, Biasucci LM, Manzoli U (1985) The efficacy of IS-5-MN in patients with ischemic heart disease: a pilot study. In: Mononitrates. Cohn JN, Ritthinghausen R (eds) Springer-Verlag, Berlin, 230–234
295. Kamphausen U (1984) Therapie mit Elantan in retardierter Form nach überstandenem Herzinfarkt. In: Mononitrat. Hochrein H, Tauchert M, Schrey A, Weiss M (eds) Universitätsdruckerei und Verlag Dr C Wolf und Sohn, München, 158–171
296. Rennhak U, Riebesel T, Biamino G (1985) A double-blind cross-over study on the effectiveness and possible development of tolerance during long-term therapy with isosorbide-5-mononitrate or isosorbide dinitrate slow-release in coronary artery disease. In: Mononitrates. Cohn JN, Rittinghausen R (eds) Springer-Verlag, Berlin, 147–153
297. Menzel M (1984) Ein-Jahresstudie mit Elantan bei Patienten mit koronarer Herzerkrankung – Wirksamkeit und Verträglichkeit im Rahmen einer Langzeitbehandlung. In: Mononitrate. Hochrein H, Tauchert M, Schrey A, Weiss M (eds) Universitätsdruckerei und Verlag Dr C Wolf und Sohn, München, 98–111

298. Heepe W (1984) Langzeittherapie der Patienten mit koronarer Herzkrankheit mit IS-5-N retard. In: Mononitrat. Hochrein H, Tauchert M, Schrey A, Weiss M (eds) Universitätsdruckerei und Verlag Dr C Wolf und Sohn, München, 172–229
299. Tauchert M, Jansen W, Osterspey A, Fuchs M, Hombach V, Hilger HH (1983) Dose dependence of tolerance during treatment with mononitrates: Z Kardiol 72 (Suppl 3):218–228
299a. Jansen W, Prenze B, Kümper H, Tauchert M (1990) Interval treatment of coronary heart disease with 50 mg sustained-release isosorbide-5-mononitrate. Am J Cardiol (in press)
300. Svendsen HH, Amtorp O (1986) A controlled, double-blind, cross-over study comparing two dose regimens of slow-release isosorbide-5-mononitrate in angina pectors. X World Congress of Cardiology, Washington, D.C., September 14–19, (795)
301. Kenedi P (1985) Intraindividuelle Dosiswirkungsbeziehung von elantan long. In: Mononitrat. Borchard U, Rafflenbeul W, Schrey A (eds) Universitätsdruckerei und Verlag Dr C Wolf und Sohn, München 108–115
302. Thadani U, Hamilton SF, Olson E et al (1987) Duration of effects and tolerance of slow-release isosorbide-5-mononitrate for angina pectoris. Am J Cardiol 59:756–762
303. Danahy DT, Aronow WS (1977) Hemodynamic and antianginal effects of high-dose oral isosorbide dinitrate after chronic use. Circulation 56:205–212
304. Eggeling T, Jansen W, Osterspey A, Tauchert M, Hilger HH (1986) Einfluß von 50 mg Isosorbid-5-Nitrat retard auf die Belastbarkeit von Patienten mit koronarer Herzkrankheit. Med Klinik 81:275–280
305. Biamino G, Oeff M, Andreson D et al (1985) Hemodynamic effects of isosorbide-5-mononitrate in patients with coronary artery disease, at rest and under exercise. In: Mononitrates. Cohn JN, Rittinghausen R (eds) Springer-Verlag, Berlin, 277–287
306. Brüggemann T, Broese K, Biamino G (1985) Assessment of changes in left ventricular performance induced by nitrates at rest using an ECG-gated scintillation probe device. In: Mononitrates. Cohn JN, Rittinghausen R (eds) Springer-Verlag, Berlin 292–296
307. Bergbauer M, Sabin G (1985) The effects of isosorbide-5-mononitrate on left ventricular function assessed by systolic time intervals. In: Mononitrates. Cohn JN, Rittinghausen R (eds) Springer-Verlag, Berlin 288–291
308. Müller G, Überbacher HJ, Glocke M (1983) Koronartherapeutische Wirksamkeit von niedrig dosiertem Isosorbid-5-Mononitrat im Vergleich zur Kombination von IS-5-MN und Metipranolol und Placebo. Med Welt 34:321–327
309. Koch G, Fransson L, Karlegärd L, Sauer W, Überbacher HJ (1985) Double-blind comparison of the effects of nifedipine and isosorbide-5-mononitrate on the circulatory response to exercise, exercise tolerance, and ST-segment depression in patients with ischemic heart disease. In: Mononitrates. Cohn JN, Rittinghausen R (eds) Springer-Verlag, Berlin, 245–253
310. Erlemeier HH, Kupper W, Bleifeld W (1985) Wirkung von Isosorbid-5-Mononitrat (IS-5-MN), Nifedipin retard und deren Kombination bei Patienten mit koronarer Herzerkrankung. In: Mononitrat. Borchard U, Rafflenbeul W, Schrey A (eds) Universitätsdruckerei und Verlag Dr C Wolf und Sohn, München, 119–124
311. Überbacher HJ, Glocke M, Abshagen U (1983) Langzeitverträglichkeit und koronartherapeutische Wirksamkeit von Isosorbid-5-Mononitrat. Münch Med Wschr 125:76–78
311a. Wisenberg G, Roks C, Nichol P, Goddard MD (1989) Sustained effect of and lack of development of tolerance to controlled-release isosorbide-5-mononitrate in chronic stable angina pectoris. Am J Cardiol 64:569–576
312. Überbacher HJ, Steinorth G, Glocke M, Abshagen U (1983) An open, long-term multicentre study of the antianginal effects and safety of isosorbide-5-mononitrate at low doses in patients with coronary heart disease. Pharmatherapeutica 3:331–341
313. Steinorth G, Seidel U, Glocke M, Überbacher HJ (1985) Double-blind, cross-over study on the antianginal efficacy of IS-5-MN and ISDN s.r., a multicenter trial. In: Mononitrates. Cohn JN, Rittinghausen R (eds) Springer-Verlag, Berlin, 270–276

314. Tauchert M, Jansen W, Ulbrich T et al (1985) Effect of nitroglycerin patches on hemodynamic and exercise parameters in patients with coronary artery disease compared with IS-5-MN and placebo. In: Mononitrates. Cohn JN, Rittinghausen R (eds) Springer-Verlag, Berlin, 265–269
315. Rabinowitz B, Hod H, Chouraqui P, Neufeld HN (1986) Comparison between the hemodynamic effects of isosorbide-5-mononitrate and isosorbide dinitrate in ischemic pump failure – an acute randomized, crossover study. X World Congress of Cardiology, Washington D.C., September 14–19 (2740)
316. Schiavoni G, Mazzari M, Montenero AS, Mongiardo R, Scabbia E, Manzoli U (1985) Hemodynamic effects of IS-5-MN in patients with congestive heart failure. In: Mononitrates. Cohn JN, Rittinghausen R (eds) Springer-Verlag, Berlin, 311–315
317. Klütsch K, Schöniger F, Koch E, Grosswendt J (1985) Der Stellenwert der Nitrate in der Therapie der chronischen Herzinsuffizienz. In: Mononitrat. Borchard U, Rafflenbeul W, Schrey A (eds) Universitätsdruckerei und Verlag Dr C Wolf und Sohn, München, 126–133
318. Tronconi L, Raisaro A, Recusani F, Gazzaniga P, Lanzarini L (1985) Efficacy of IS-5-MN (ISMO 20) in patients with coronary artery disease and impaired left-ventricular function. In: Mononitrates. Cohn JN, Rittinghausen R (eds) Springer-Verlag, Berlin, 306–310
319. Keck M (1984) Monotherapie der chronischen Linksherzinsuffizienz mit IS-5-MN. In: Mononitrat. Hochrein H, Tauchert M, Schrey A, Weiss M (eds) Universitätsdruckerei und Verlag Dr C Wolf und Sohn, München, 88–96
320. VandenBurg MJ, Marks C, Woodings D, Stephens D (1986) Lack of tolerance to oral isosorbide-5-mononitrate after four weeks' therapy. X World Congress of Cardiology, Washington, D.C., September 14–19 (1101)
321. Gammage MD, Murray RG, Littler WA (1986) Isosorbide mononitrate by intravenous bolus administration in acute left ventricular failure. X World Congress of Cardiology, Washington, D.C., September 14–19 (2158)
322. Pozzoli G, Gandini R, Baggio E, Montella S, Vargiu B, Cunietti E (1986) Effect of isosorbide-5-mononitrate on exercise capacity in patients with congestive heart failure. X World Congress of Cardiology, Washington, D.C., September 14–19 (2730)
323. Rabinowitz B (1985) Oral isosorbide-5-mononitrate and isosorbide dinitrate in ischemic pump failure (acute hemodynamic cross-over study). In: Mononitrat. Borchard U, Rafflenbeul W, Schrey A (eds) Universitätsdruckerei und Verlag Dr C Wolf und Sohn, München, 89–93
324. Tanaka K, Takada K, Seino Y, Katoh T, Takano T (1986) Hemodynamic effects and pharmacokinetics of isosorbide-5-mononitrate (ISMN) on congestive heart failure – comparative study with ISDN. X World Congress of Cardiology, Washington, D.C., September 14–19 (2920)
324a. Bennet D, Fitzgerald L, Lis Y, Bland M (1987) Possible benefits of early isosorbide-5-mononitrate therapy in acute myocardial infarction. Eur Heart J 8 (Suppl I):37
325. Hollenberg M, Go M (1984) Clinical studies with transdermal nitroglycerin. Am Heart J 108 (1):223–231
326. Crean PA, Ribeiro P, Crea F, Davies GJ, Ratcliffe D, Maseri A (1984) Failure of transdermal nitroglycerin to improve chronic stable angina: a randomizedm placebo-controlled, double-blind, crossover trial. Am Heart J 108:1494–1500
327. Olivari MT, Carlyle PF, Levine B, Cohn JN (1983) Hemodynamic and hormonal response to transdermal nitroglycerin in normal subjects and in patients with congestive heart failure. J Am Coll Cardiol 2:872–878
328. Reichek N, Priest C, Zimrin D, Chandler T, Sutton MSJ (1984) Clinical antianginal effects of nitroglycerin patches. Am J Cardiol 54:1–7
329. Rajfer SL, Demma FJ, Goldberg LI (1984) Sustained beneficial hemodynamic responses to large doses of transdermal nitroglycerin in congestive heart failure and comparison with intravenous nitroglycerin. Am J Cardiol 54:120–125
330. Abrams J (1984) The brief saga of transdermal nitroglycerin discs: paradise lost? Am J Cardiol 54:220–224

331. Schneeweiss A (1986) Nitroglycerin. In: Drug Therapy in Cardiovascular Diseases. Lea & Febiger, Philadelphia, 5
332. Greco R, D'Alterio D, Schiattarella M, Boccia A, Greco L, Marsico F (1988) Efficacy of a new transdermal nitroglycerin patch (deponit 10) in patients with stable angina pectoris. Am J Cardiol (Suppl) 61 (9):44E–51E
333. Rezakovic DE (1985) The use of nitroglycerin patches over 6 months to treat patients with stable angina pectoris. Herz/Kreislauf (Suppl):19–21
333a. Rezakovic DE, Pavicic L, Stalec J (1987) Belastungskapazität während einer 12monatigen Therapie der stabilen Angina pectoris mit transdermalem Nitroglycerin. Herz/Kreislauf (Suppl I):34–37
334. Schiavoni G, Mazzari M, Lanza G, Frustaci A, Mongiardo R, Pennestri F (1982) Evaluation of the efficacy and the length of action of a new preparation of slow-release nitroglycerin for percutaneous absorption (Nitro-Dur, Sigma-Tau) in angina pectoris caused by exercise. J Clin Pharmacol Res 2 (Suppl 1):15
335. Georgopoulos AJ, Markis A, Georgiadis H (1982) Therapeutic efficacy of a new transdermal system containing nitroglycerin in patients with angina pectoris. Eur J Clin Pharmacol 22:481–485
336. Thompson R (1983) Clinical use of transdermal delivery devices with nitroglycerin. Circulation 67 (Suppl III):III–407
336a. Greco R, Schiattarella M, Wolff S, D'Alterio D, Tartaglia P (1990) Long-term efficacy of nitroglycerin patch in stable angina pectoris. Am J Cardiol (in press)
337. Atkins JM (1986) Some issues concerning transdermal nitroglycerin patches. Am Heart J 112:229–232
338. Müller P, Imhof PR, Burkart F, Chu LC, Gerardin A (1982) Human pharmacological studies of a new transdermal system containing nitroglycerin. Eur J Clin Pharmacol 22:473–480
339. Imhof PR, Müller P, Georgopoulos AJ, Garnier B (1985) Nitroderm TTS versus oral isosorbide dinitrate. A double blind trial in patients with angina pectoris. Acta Ther 11:155
340. Pucci P, Zambaldi G, Cerisano G, Roccanti P (1983) Evaluation of a new preparation of transdermal nitroglycerin for patients with angina of effort. Ital Cardiol 13:167–171
341. Wester HA, Mouselimis N (1984) Therapeutic efficacy of a transdermal system containing nitroglycerin in comparison with nifedipine. Z Kardiol 73:510–514
342. Fazi M, Novo S, Giannola A, Adamo L, Alaimo G, Strano A (1984) Effects of transdermal application of trinitroglycerin on exercise performance and thallium stress scintigraphy in patients with effort angina pectoris. Eur Heart J 5 (Suppl 1):269 (1368) abstr
343. James MA, Walker PR, Papouchado M, Wilkinson PR (1985) Efficacy of transdermal glyceryl trinitrate in the treatment of chronic stable angina pectoris. Br Heart J 53:631–635
344. Muiesan G, Agabiti-Rosei E, Muiesan L et al (1986) A multicenter trial of transdermal nitroglycerin in exercise-induced angina: individual antianginal response after repeated administration. Am Heart J 112:233–238
345. Charash B, Scheidt SS (1986) The controversy over transdermal nitroglycerin: an update. Am Heart J 112:207–215
346. Martines C (1984) Comparison of the prophylactic antianginal effect of two doses of nitroderm TTS in out-patients with stable angina pectoris. Curr Ther Res 36:483–489
347. Schneider W, Michel O, Kaltenbach M, Bussmann WD (1985) The antianginal effect of transdermally applied nitroglycerin, as a function of the patch size. Dtsch Med Wschr 110:87–91
348. Rezakovic-Music DE (1987) Controversies on drug tolerance after chronic application of NTG patch. Eur Heart J 8 (Suppl 1):17
348a. Schaer DH, Buff LA, Katz RJ (1988) Sustained antianginal efficacy of transdermal nitroglycerin patches using an overnight 10-hour nitrate-free interval. Am J Cardiol 61:46–50

348b. Cowan JC, Bourke JP, Reid DS, Julian DG (1987) Prevention of tolerance to nitroglycerin patches by overnight removal. Am J Cardiol 60:271–275
348c. Luke R, Sharpe N, Coxon R (1987) Transdermal nitroglycerin in angina pectoris: efficacy of intermittent application. J Am Coll Cardiol 10:642–646
348d. Reiniger G, Lehmann G, Rudolph W (1989) Persisent anti-ischemic effects of NTG-patches through interval treatment with increasing nitroglycerin plasma levels. Eur Heart J 10 (Suppl):74 (358) abstr
348e. DeMots H, Glasser SP (1989) Intermittent transdermal nitroglycerin therapy in the treatment of chronic stable angina. J Am Coll Cardiol 13:786–793
348f. The Dutch Nitro Study Group (1989) Long term efficacy of continuous and intermittent use of transdermal nitroglycerin patches in angina pectoris. Eur Heart J 10 (Suppl):73 (p352) abstr
348g. Nabel EG, Barry J, Rocco MB, Mead K, Selwyn AP (1989) Effects of dosing intervals on the development of tolerance to high dose transdermal nitroglycerin. Am J Cardiol 63:663–669
349. Flaherty JT (1986) Usefulness of intravenous and transdermal nitroglycerin therapy in patients with unstable angina. Am Heart J 112:220–221
350. Shell WE, Kivowitz CF, Rubins SB, See J (1986) Mechanisms and therapy of silent myocardial ischemia: the effect of transdermal nitroglycerin. Am Heart J 112:222–229
351. Olivari MI, Carlyle PF, Levine B, Cohn N (1983) Hemodynamic and hormonal response to transdermal nitroglycerin in normal subjects and in patients with congestive heart failure. J Am Coll Cardiol 2:872–878
352. Sharpe DN, Coxon R (1984) Nitroglycerin in a transdermal therapeutic system in chronic heart failure. J Cardiovasc Pharmacol 6:76–82
353. Armstrong PW (1987) Pharmacokinetic-hemodynamic studies of transdermal nitroglycerin in congestive heart failure. J Am Coll Cardiol 9:420–425
354. Elkayam U, Kulick D, McIntosh N, Roth A, Rahimtoola SH (1987) Incidence of early tolerance to continuos intravenous nitroglycerin infusion in patients with coronary artery disease and heart failure. J Am Coll Cardiol 9:103A abstr
355. Cowan C, Bourke J, Reid DS, Julian DG (1986) Prevention of tolerance to nitroglycerin patches by intermittent dosing. Circulation 74:II–137 (abstr)
356. Blasini R, Froer Kl, Blümel G, Rudolph W (1982) Wirkungsverlust von Isosorbiddinitrat bei Langzeitbehandlung der chronischen Herzinsuffizienz. Herz 7:250–258
357. Schneeweiss A (1986) Isosorbide dinitrate. In: Drug Therapy in Cardiovascular Diseases. Lea & Febiger, Philadelphia, 29
357a. Sharpe N, Coxon R, Webster M, Luke R (1987) Hemodynamic effects of intermittent transdermal nitroglycerin in chronic congestive heart failure. Am J Cardiol 59:895–899
358. Nelson GIC, Silke B, Verma SP, Abdulali S, Clarke J, Taylor SH (1984) Transdermal nitroglycerin in acut LV failure following myocardial infarction (M.I.). Eur Heart J 5 (Suppl 1):269 (1369)
359. Roth A, Freidenberger L, Baruch T, Rahimtoola SH, Elkayam U (1986) Hemodynamic effect of nitrate therapy during isometric exercise in patients with chronic heart failure. J Am Coll Cardiol 7:233A abstr
360. Bennett ED, Davis AL (1984) A haemodynamic and pharmacokinets study to assess a new transdermal nitroglycerin preparation in normal subjects. Eur J Clin Pharmacol 26:293–296
361. Cohn JN (1985) Nitrate therapy of congestive heart failure. In: Mononitrates. Cohn JN, Rittinghausen R (eds) Springer-Verlag, Berlin, 299–305
362. Klein HO, Ninio R, Blank I et al (1983) Prolonged (24-hour or more) hemodynamic effect of a slow-realease nitroglycerin ointment. A radionuclide study. Circulation 67 (Suppl III):III–406
363. Armstrong PW, Moffat JA (1983) Tolerance to organic nitrates: clinical and experimental perspektives. Am J Med 74 (6 B):73–84
364. Fung H-L (1984) Pharmacokinetic determinants of nitrate action. Am J Med 76/6A:22–26

365. Barclay SA, Bennett D, Adam J, Valantine H, Boyle GJ, Mannering D (1986) Plasma volume expansion may explain tolerance to transdermal nitroglycerin. Circulation 74 (Suppl II):II–321 (1279)
366. Zwicke DL, Niazi I, Reeves WC, Wagle SS (1986) Reduced transcutaneous nitroglycerin absorption in blacks. Circulation 74 (Suppl II):II–136 (543)
367. Barkve TF, Langseth-Manrique K, Bredesen JE, Gjesdal K (1986) Increased uptake of transdermal glyceryl trinitrate during physical exercise and during high ambient temperature. Circulation 74 (Suppl II):II–136 (542)
368. Sullivan M, Savvides M, Abouantoun S, Birk-Madsen E, Froelicher V (1985) Failure of transdermal nitroglycerin to improve exercise capacity in patients with angina pectoris. J Am Coll Cardiol 5:1220–1223
369. Imhof PR, Ott B, Weiss A, Chu LC, Chasseaud LF (1981) Plasma concentrations and hemodynamic effects of percutaneously administered nitroglycerin and isosorbide dinitrate in healthy volunteers. In: Nitrates III. Lichtlen PR, Engel H-J, Schrey A, Swan HJC (eds) Springer-Verlag, Berlin, 66–74
370. Handler CE, Sullivan I (1984) Double blind crossover trial comparing sustained release isosorbide dinitrate cream and tablets in stable angina. Eur Heart J 5 (Suppl 1):34
370a. Kiowski W, Schwendener R, Bassignana V, Pfisterer M, Burkart F (1989) Can isosorbidedinitrate-tolerance be prevented by once daily administration? Eur Heart J 10 (Suppl):74 (p355) abstr
371. Windorfer A (1982) Nitroglycerin blood level after buccal and oral application of synchron nitroglycerin. In: Controlled Release Nitroglycerin in Buccal and Oral Form. Advances in Pharmacotherapy, vol 1. Bussmann WD et al (eds) Basel, S Karger, 44–48
372. Bussmann WD, Kromm W, Reifart N, Fung HL (1982) Acute hemodynamic and pharmacokinetic effects of buccal nitroglycerin. In: Controlled Release Nitroglycerin in Buccal and Oral Form. Advances in Pharmacotherapy, vol 1. Bussmann WD et al (eds), Basel, S Karger, 64–72
373. Langbehn AF, Beurich HW, Leupold H, Hoffmann R, Lohaus R (1982) Effect of buccal synchron nitroglycerin on pulmonary artery pressure at rest and during exercise. In: Controlled Release Nitroglycerin in Buccal and Oral Form. Advances in Pharmacotherapy, vol 1. Bussmann WD et al (eds) Basel, S Karger, 73–97
374. Abrams J (1983) New nitrate delivery systems: buccal nitroglycerin. Am Heart J 105:848–854
375. Reichek N, Priest C, Kienzle RN, Kleaveland JP, Bolton S, Sutton MStJ (1982) Angina prophylaxis with buccal synchron nitroglycerin: a rapid onset long-acting nitrate. In: Controlled Release Nitroglycerin in Buccal and Oral Form. Advances in Pharmacotherapy, vol 1. Bussmann WD et al (eds) Basel, S Karger, 143–154
376. Lahiri A, Sonecha TN, Crawley JW, Raftery EB (1982) Efficacy of sustained release buccal nitroglycerin in patients with ischaemic cardiomyopathy. Clin Sci 63:38 (104) abstr
377. Abrams J (1983) Patient acceptability of buccal nitroglycerin – a multicenter study. In: Goldberg AA, Parsons DG (eds) Modern Concepts in Nitrate Delivery Systems. Int. Congress and Symposium Ser (54) Royal Society of Medicine, Adacemic Press London 151–159
378. Wei JY, Reid PR (1979) Quantitative determination of trinitroglycerin in human plasma. Circulation 59:588–592
379. Yap SK, McNiff EF, Fung H-L (1978) Improved GLC determination of plasma nitroglycerin concentrations. J Pharm Sci 67:582–584
380. Rosseel MT, Bogaert MG (1973) GLC determination of nitroglycerin and isosorbide dinitrate in human plasma. J Pharm Sci 62:754–758
381. Wei JY, Reid PR (1981) Relation of time course of plasma nitroglycerin levels to echocardiographic, arterial pressure and heart rate changes after sublingual administration of nitroglycerin. Am J Cardiol 48:778–782
382. Zimrin D, Reichek N, Bogin K, Cameron S, Douglas P, Fung H-L (1986) I.V. Nitroglycerin: Infusion rate, plasma concentration and hemodynamic effects. J Am Coll Cardiol 7:181A abstr

383. McNiff EF, Yap PSK, Fung H-L (1980) Nitroglycerin. In: Analytical Profiles of Drug Substances, vol 9. Florey (ed) Academic Press, New York
384. Straehl S, Galeazzi RL (1985) Isosorbide dinitrate bioavailability, kinetics and metabolism. Clin Pharmacol Ther 38:140–149
385. Assinder DF, Chasseaud LF, Taylor T (1977) Plasma isosorbide dinitrate concentrations in human subjects after administration of standard and sustained release formulations. J Pharm Sci 66:775–778
386. Fung H-L, McNiff EF, Ruggirello D, Darke A, Thadani U, Parker JO (1981) Kinetics of isosorbide dinitrate and relationship to pharmacological effects. Br J Clin Pharmacol 11:579–590
387. Chasseaud LF (1983) Newer aspects of the pharmacokinetics of organic nitrates. Z Kardiol 72 (Suppl 3):20–23
388. Taylor T, Chasseaud LF, Doyle E, Bonn R, Darragh A, Lambe RF (1982) Isosorbide dinitrate pharmacokinetics. Arzneim Forsch 32:1329–1333
389. Chasseaud LF, Down WH, Sacharin RM (1978) Effect of repeated oral administration of isosorbide dinitrate on hepatic glutathione S-transferase activity in the rat. Biochem Pharmacol 27:1695–1697
390. Fung H-L, Ruggirello D, Stone JA, Parker JO (1983) Effects of disease, route of administration, cigarette smoking, food intake on the pharmacokinetics and circulatory effects of isosorbide dinitrate. Z Kardiol 72 (Suppl 3):5–10
391. Bogaert MG (1983) Clinical pharmacokinetics of organic nitrates. Clin Pharmacokinetic 8:410–421
392. Bogaert MG, Rosseel MT, Elewaut A (1984) Fate of orally given isosorbide dinitrate in cirrhotic patients. Int J Clin Pharmacol, Ther Toxicol 22:491–492
393. Abshagen U (1985) Pharmacokinetics of ISDN, sustained-release ISDN and IS-5-MN. In: Mononitrates. Cohn JN, Rittinghausen R (eds) Springer-Verlag, Berlin, 53–66
394. Straehl P, Galeazzi RL, Soliva M (1984) Isosorbide-5-mononitrate and isosorbide-2-mononitrate kinetics after intravenous and oral dosing. Clin Pharmacol Ther 36:485–492
395. Taylor T, Chasseaud LF, Major R, Doyle E (1981) Isosorbide-5-mononitrate pharmacokinetics in human. Biopharm Drug Dispos 2:255–263
396. Reifart N, Bussmann WD, Schirmer M, Kaltenbach M (1981) Hämodynamische Wirksamkeit, Wirkdauer und Pharmakokinetik von 80 mg Isosorbid-5-Mononitrat beim frischen Herzinfarkt. Med Welt 32:540–542
397. Bartsch W, Strein K, Bohm E, Sponer G, Müller-Beckmann B (1985) Isosorbide-5-mononitrate increases cyclic guanosine monophosphate concentration in rat aorta in vitro and in vivo. In: Mononitrates. Cohn JN, Rittinghausen R (eds) Springer-Verlag, Berlin, 34–38
398. Akpan W, Endele R, Neugebauer G, Steudel H (1985) Pharmacokinetics of IS-5-MN after oral and intravenous administration in patients with hepatic failure. In: Mononitrates. Cohn JN, Rittinghausen R (eds) Springer-Verlag, Berlin, 86–91
399. Mannebach H, Ohlmeier H, Möllendorff E, Gleichmann U, Abshagen U (1981) Steady-state Kinetik von Isosorbid-5-Mononitrat bei Patienten mit koronarer Herzkrankheit. Med Welt 32:517–520
400. Raue F, Ritz E, Akpan W, Endele R, Neugebauer G (1985) Kinetics of IS-5-MN and its glucuronide in patients with renal failure. In: Mononitrates. Cohn JN, Rittinghausen R (eds) Springer-Verlag, Berlin, 78–85
401. Davies DE, Williams PEO (1985) Pharmacokinetics and relative bioavailability of isosorbide-5-mononitrate after chronic oral dose administration of 2 formulations (elantan tablets vs elantan long capsules). In: Mononitrat. Borchard U, Raffleneul W, Schrey A (eds) Universitätsdruckerei und Verlag Dr C Wolf und Sohn, München, 13–20
401a. Evers J, Bonn R, Boertz A, Cawello W, Dickmans HA, Weiss M (1989) Pharmacokinetics of isosorbide dinitrate, isosorbide-2-nitrate and isosorbide-5-nitrate in renal insufficiency after repeated oral dosage. Klin Wochenschr 6:342–348

402. Strein K, Sponer G, Bartsch W (1985) Differences in the metabolism of isosorbide-5-mononitrate and isosorbide dinitrate and their consequences for toxicologic findings. In: Mononitrates. Cohn JN, Rittinghausen R (eds) Springer-Verlag, Berlin, 45–49
403. Fung H-L, Sutton SC, Kamiya A (1984) Blood vessel uptake and metabolism of organic nitrates in the rat. J Pharmacol Exp Ther 228:334–341
404. Imhof PR, Ott B, Frankhauser P, Chu L-C, Hodler J (1980) Difference in nitroglycerin dose-response in the venous and arterial beds. Eur J Clin Pharmacol 18:455–460
405. Imhof PR, Vuillemin TH, Gérardin A, Racine A, Müller P, Follath F (1984) Studies of the bioavailability of nitroglycerin from a transdermal therapeutic system (Nitroderm TTS). Eur J Clin Pharmacol 27:7–12
406. Distante A, Lombardi M, Moscarelli E, Morales MA, L'Abbate A (1986) The therapeutic role of isosorbide-5-mononitrate in stable and unstable angina pectoris. Z Kardiol 75 (Suppl 3):38–41
407. Schneider W, Stahl B, Kaltenbach M, Bussmann WD (1982) Dose effect relationship of isosorbide dinitrate in the treatment of angina pectoris. Dtsch Med Wschr 107:771–776
407a. Wolf H, Reiniger G, Lehmann G, Dirschinger J, Rudolph W (1989) Evidence for rebound-phenomena after withdrawal of nitroglycerin. Eur Heart J 10 (Suppl):74 (p357) abstr
408. Franciosa JA, Blank RC, Cohn J (1978) Nitrate effects on cardiac output and left ventricular outflow resistance in chronic congestive heart failure. Am J Med 64:207–213
408a. Ferguson JJ, Diver DJ, Boldt M, Paternak RC (1989) Significance of nitroglycerin-induced hypotension with inferior wall acute myocardial infarction. Am J Cardiol 64:311–314
408b. Figueras J, Moya A, Monné J (1989) Blood volume and right heart hemodynamics in abrupt hypotension following sublingual isosorbide dinitrate in acute myocardial infarction. Eur Heart J 10 (3):244–249
409. Nemerovski M, Shah PK (1981) Syndrome of severe bradycardia and hypotension following sublingual nitroglycerin administration. Cardiology 67:180–189
410. Come PC, Pitt B (1986) Nitroglycerin-induced severe hypotension and bradycardia in patients with acute myocardial infarction. Circulation 54:624–628
411. von Arnim T, Autenrieth G, Bolte H-D (1984) Acute myocardial infarction during continuous electrocardiographic ST-segment recording. Possible role of bradycardia and hypotension induced by glyceryl trinitrate. Br Heart J 51:575–577
412. Prodger SH, Ayman D (1932) Harmful effects of nitroglycerin. With special reference to coronary thrombosis. Am J Med Sci 184:480–491
413. Weissler AM, Warren JV, Estes EH, McIntosh HD, Leonard J (1957) Vasodepressor syncope: Factors influencing cardiac output. Circulation 15:875–882
414. Karp HR, Weissler AM, Heyman A (1961) Vasodepressor syncope: EEG and circulatory changes. Arch Neurol 5:94–101
415. Arsura E, Guadagnino V, Lichstein E et al (1982) Methemoglobin levels produced by commonly used nitrates. Circulation 66 (Suppl II):II–309 (1236) abstr
416. Gibson GR, Hunter JB, Raabe DS Jr, Manjoney DI, Ittleman FP (1982) Methemoglobinemia produced by high-dose intravenous nitroglycerin. Ann Intern Med 96:615–616
417. Daskoff N, Roland JMA, Varghese PJ, Pitt B (1976) Effect of nitroglycerin on ventricular fibrillation threshold of nonischemic myocardium. Am J Cardiol 38:184–188
418. Glisson SN, El-Etr AA, Lim R (1979) Prolongation of pancuronium-induced neuromuscular blockade by intravenous infusion of nitroglycerin. Anesthesiology 51:47–49
419. Rogers MC, Traystman RJ, Epstein MH (1979) Nitroglycerin effects on intracranial pressure. Am J Cardiol 43:342-abstr
420. Shook TL, Kirshenbaum JM, Hundley RF, Shorey JM, Lamas GA (1984) Ethanol intoxication complicating intravenous nitroglycerin therapy. Ann Intern Med 101:498–499
421. Shorey I, Bhardwaj N, Loscalzo J (1984) Acute Wernicke's encephalopathy after intravenous infusion of high-dose nitroglycerin. Ann Intern Med 101:500
422. Weber S, Rey E, Pipeau C et al (1983) Influence of aspirin on the hemodynamic effects of sublingual nitroglycerin. J Cardiovasc Pharmacol 5:874–877

423. Simonetti I, de Caterina R, Marzilli M, de Nes M, L'Abbate A (1983) Coronary vasodilation by nitrates is not mediated by the prostaglandin system: an angiographic and hemodynamic study. Z Kardiol 72 (Suppl 3):40–45
424. Van Dusen J, Fischi J (1981) Inhibition of nitroglycerin effect in humans by suppression of prostaglandin E. Am J Cardiol 47:390-abstr
425. Thadani U, Kellermann D (1983) Interaction of indomethacin and nitroglycerin on hemodynamics and exercise tolerance in patients with angina pectoris. Z Kardiol 72 (Suppl 3):35–39
426. Hyldstrup L, Christiansen C, Nielsen MD, Transbol I (1983) Orthostatic response before and after nitroglycerin in metropolol- and verapamil-treated angina pectoris. Acta Med Scand 214:131–134
427. Habbab MA, Haft JI (1986) Heparin resistance induced by intravenous nitroglycerin. Circulation 74 (Suppl II):II–321 (1280) abstr
427a. Pizzulli L, Nitsch J, Lüderitz B (1989) Nitroglycerin inhibition of the heparin effect. Eur Heart J 10 (Suppl):116 (582) abstr
428. Tauchert M, Jansen W, Ulbrich T, Meyer L (1984) Dosisabhängige Veränderungen hämodynamischer Parameter unter Isosorbid-5-Nitrat. In: Mononitrat. Hochrein H, Tauchert M, Schrey A, Weiss M (eds) Universitätsdruckerei und Verlag Dr C Wolf und Sohn, München, 67–73
429. Jansen W, Tauchert M, Osterspey A, Höpp HW, Wilcken K (1984) Systemische und koronare Hämodynamik bei Elantan 20 und 50. In: Mononitrat. Hochrein H, Tauchert M, Schrey A, Weiss M (eds) Universitätsdruckerei und Verlag Dr C Wolf und Sohn, München, 38–58
430. Wenger NK, Hellerstein HK, Blackburn H, Castronova SJ (1982) Physician practice in the management of patients with uncomplicated myocardial infarction: changes in the past decade. Circulation 65:421–427
431. Dalen JE, Goldberg RJ, Gore JM, Struckus J (1984) Therapeutic interventions in acute myocardial infarction. Chest 86:257–262
432. Maseri A (1985) A review of nitrate therapy in stable angina, variant angina, unstable angina and myocardial infarction. Z Kardiol 74 (Suppl 4):1–3

FSC
www.fsc.org
MIX
Papier | Fördert gute Waldnutzung
FSC® C083411

Kolibri 360 GmbH
Ferdinand-Jühlke-Straße 7
99095 Erfurt, Deutschland
produktsicherheit@kolibri360.de